EXTERNAL PRESSURE, NATIONAL RESPONSE

Industrial Adjustment in Canada since the 1970s

Prosper M. Bernard, Jr.

University Press of America,® Inc.
Lanham · Boulder · New York · Toronto · Plymouth, UK

Copyright © 2009 by
University Press of America,® Inc.
4501 Forbes Boulevard
Suite 200
Lanham, Maryland 20706
UPA Acquisitions Department (301) 459-3366

Estover Road
Plymouth PL6 7PY
United Kingdom

Library of Congress Control Number: 2009922983
ISBN: 978-0-7618-4578-2 (paperback : alk. paper)
eISBN: 978-0-7618-4579-9

To my parents

Prosper and Francine Bernard

Contents

Acknowledgments

This book originated as a Ph.D. dissertation at the Graduate School of the City University of New York. I want to thank John Bowman, Kenneth Erickson, Howard Lentner, Irving Leonard Markovitz, and W. Ofuatey-Kodjoe for providing me with valuable comments. I also wish to thank the many people who helped me during my field research at McGill University and Concordia University in Montreal, and at Carleton University and the National Library of Canada in Ottawa. In particular, I am grateful to Jean-Philippe Côté, Yves Bélair, and Michel Plaisent for their help during my research.

Different parts of this book were presented at the 2004 Annual Meeting of the Canadian Political Science Association, the 2005 Annual Meeting of the Midwest Political Science Association, and the 2005 Biennial Meeting of the Association for Canadian Studies in the United States. I thank the participants in these presentations for providing me with helpful comments.

I owe a considerable debt to Howard Lentner—my Ph.D. advisor, a colleague at Baruch College, and a friend. Howard's generosity, encouragement, and guidance were instrumental at every stage of this project. I would like to thank Ross Burkhart and Bill Raynor for reading and commenting on many parts of the manuscript. Roland Perron deserves special mention for his fine editorial work.

Above all, I would like to thank my family. My brothers, Erik and Jean-Sébastien, have always been a constant source of encouragement in their unique ways. I would have never opted for a career in the academic world had it not been for my parents, Francine and Prosper Bernard. They have provided me with a lifetime of unconditional support and love. It is to them that this book is dedicated.

Chapter One

Introduction

The industrial adjustment challenges facing Canada starting in the late 1960s were linked to the economic and political effects of its economic relationship with the United States. The adjustment problem assumed three interlocking characteristics. First, Canada's share of world manufacturing activities dropped during the 1960s and early 1970s, which raised concerns about the economy's capacity to compete internationally. Second, the Canadian economy was more and more penetrated by American capital and this, consequently, reduced Canada's leverage over its own course of industrial development. Finally, the increasingly aggressive economic policies of the United States starting in the 1960s, illustrated most vividly with the 1971 Nixon measures and capricious use of trade remedy law, heightened Canada's sense of vulnerability and raised fresh worries over its ability to maintain policy autonomy.

In this book I argue that the sequence of adjustment actions reflected policymakers' responses to changes and continuities in the domestic and international constraints and incentives facing them. At the international level, I argue that increases in Canada's vulnerability dependence vis-à-vis the United States and changes in the asymmetry of economic power between Canada and the United States defined the external matrix of obstacles and opportunities for state action. Whereas the decline in Canada's relative economic standing and moderate level of vulnerability dependence in the late 1960s and early 1970s encouraged government officials to pursue a strategy of economic nationalism in the 1970s and early 1980s, the improvements in Canada's relative standing and higher level of vulnerability dependence beginning in the late 1970s led Ottawa to shift to a strategy of liberal continentalism starting in the mid-1980s.

The sequence of adjustment actions was also shaped by domestic institutional forces. Institutional factors intervened strongly as policy entrepreneurs sought to implement their adjustment strategies. Officials were unsuccessful at implementing the strategy of economic nationalism because they were unable to mobilize sufficient social and political support to reinforce and entrench the new governmental institutions being created to administer the new strategy. As a result, an incoherent policy regime emerged, in which the strategy's policy goals were incompatible with the capacities of the state. On the other hand, the strategy of liberal continentalism was successfully implemented. The fact that this alternative policy solution to Canada's adjustment problems enjoyed broad societal and state support and that its corresponding institutional reforms matched well with the preexisting institutional structure, helped render coherent this new policy regime, thus assuring its institutionalization.

National Adjustment to International Change

States constantly adjust to international change. "Of the many international and domestic forces that set states in motion," as Ikenberry (1986, 56) points out, "none is more important than the constant pressure for national adjustment to international change produced by constant differential change between national and international systems." States respond to international change because it affects national wealth and power, as well as states' ability to maintain domestic welfare and political independence. Moreover, international economic or political change affects states differently—it creates relative winners and losers. As Gilpin (2001, 80) notes:

> Modern nation-states are extremely concerned about the consequences of international economic activities for the distribution of economic gains. Over time, the unequal distribution of gains will inevitably change the international balance of economic power, and will thus affect national security. For this reason, states have always been very sensitive to the effects of the international economy on relative rates of economic growth.

Wealth and power form two sides of the same coin. Because wealth is instrumental in projecting power and influence and because both of these are important to accomplish other state goals, states will seek to maximize their wealth. However, because states are also concerned with their relative position in the international system, they will assess their capabilities in relative terms— that is, how economic gains are divided among states. In the event that market outcomes become unfavorable to a state, that state will seek to change the current pattern of wealth distribution. It will do so in order to bring about a more favorable distributional outcome that in turn preserves and improves its relative power even though it was gaining before in absolute terms. The purpose of economic adjustment is to reshape the interplay between national power and wealth so that it generates conditions favorable to the attainment of state interests.

States that fail to respond effectively to international change can face deepening economic dependence. In its most fundamental form, dependence refers to an asymmetric power relationship between two states. When dependence is not countered, states run the risk of losing the capacity to influence the rate and direction of national economic development and to maintain political autonomy. Notwithstanding the fact that some states react more promptly than others, states always respond to the problem of dependence for the same reason—namely, to thwart the liability of vulnerability and external domination.

National adjustment can be undertaken by private actors or state officials. Heightened international competition caused by changes in international market conditions can force firms to rationalize their operations, seek joint ventures or mergers with local or international firms, engage in selective divestiture, diversify into new lines of production, augment foreign direct investment activities, or perform a combination of the above options. In addition, firms may seek political solutions to deal with international economic change. Threatened industries may receive government support, especially if the government recognizes them as being important for the wealth, industrial capacity, and employment level of the nation. Subsidies, bailouts, import protection, export promotion, among other compensatory and stabilization instruments, can be employed to shore up industries.

In this book, I will explore adjustment measures undertaken by the state. Although private-initiated and state-initiated adjustment may seek to accomplish a common objective, the two sets of actors may not always define the adjustment problem the same way, nor coordinate their actions even though they agree on the nature of the economic adjustment problem. For example, a high level of foreign participation in a national economy is more likely to raise greater concern for state officials than for the business community because the state is preoccupied not just with the performance of the national economy, but also with the political consequences of a market economy, particularly the prospect of economic vulnerability and the erosion of political independence. At other times, although the private sector and the state are in agreement with the diagnosis of an adjustment problem, there may be a lack of coordination because the two endorse different policy ideas.

An industrial adjustment strategy refers to a set of coherent policies that are derived from a common body of economic ideas. Nationally, such a strategy is formulated and implemented by central executive officials over a period of time and aims to achieve the necessary national industrial adaptation to deal with adverse external and domestic changes. Adjustment strategies are steeped in politics. Political conflicts within and between society and the state are bound to occur because of the distributional impact of such strategies. Moreover, adjustment strategies require government officials to mobilize the cooperation of various groups in society. Internationally, such strategies may seek to shift the burden of adjustment to other countries as many powerful states have attempted to do. Or, they may attempt to obtain the cooperation of other countries in order to arrive at a mutually beneficial multilateral solution to the problem of adjustment, as less powerful states have sometimes tried to do.

When dealing with economic adjustment challenges, a state will act strategically within a set of international and domestic constraints and incentives. Adjustment policies are rarely made and executed on the basis of functional efficiency. Instead, policymakers address the adjustment problem by maneuvering within a labyrinth of obstacles, laboring under the weight of their organizational and instrumental limitations, and often proceeding in accordance with a trial-and-error and bounded rationality process of decision-making.

Adjustment strategies generally aim to achieve some combination of absolute and relative gains. They can be designed to promote positive change by exploiting economies of scale and existing comparative advantages, or by fostering new ones altogether. The result of pursuing absolute gains is that an economy may be able to expand the scope of industrial activities, improve overall productivity, boost economic growth, and create more high-skilled employment opportunities. At the international level, the absolute gains function of adjustment strategies can take the form of expanding and deepening economic links with other countries. The relative gains aspect of adjustment strategies, on the other hand, focuses on the allocation of gains derived from existing patterns of economic activities—that is, it seeks to improve the relative achievement of gains by rectifying adverse distributive patterns. The purpose is not so much to create new wealth as to secure a more favorable division of gains for a dissatisfied state.

Domestic and international spheres serve as two distinct but interconnected arenas for problem solving. Strategies generally work their effects at both levels, but they tend to be designed to take advantage of an arena in which state capacity is most pronounced. The balance between a state's domestic and international capacities figures largely in determining the general orientation of an adjustment strategy. As Krasner (1985, 11) points out:

> Weakness stems from the inability to influence unilaterally or to adjust internally to the pressures of global markets. Larger industrialized states are able to influence the international environment and can adjust internally. Smaller industrialized states have little influence over the global pattern of transactions, but their domestic political economies allow them to adjust. Small size and inflexible domestic structures make Third World states vulnerable.

In this book, reference to Canada's domestic capacity refers to the federal state's ability to mobilize public support for the purpose of achieving economic adjustment goals. It also refers to the ability to rework domestic institutions so that they match the tasks of adjustment policies. The international capacity of the Canadian state refers to its ability to wield influence in international affairs in a way that advances and preserves state interests. Internationally weak states tend to use domestic policies to address the adjustment problem, whereas internationally strong but domestically weak states tend to address the adjustment problem using international instruments. Nevertheless, even in the arena in which they are relatively weak, states will labor in a trial-and-error manner to find the most suitable course of action to address the adjustment problem.

The origin of Canada's economic adjustment problem in the late 1960s was the differential rate of economic progress of Canada relative to other advanced industrial economies. The prevailing economic conditions undermined Canada's relative standing and prompted concern about the country's ability to advance and protect its national interests. In response to its adjustment problem, Ottawa employed two industrial adjustment strategies over the next three decades: economic nationalism, which was the course of action pursued by the federal state throughout the 1970s and early 1980s, and liberal continentalism, the strategy that Canada followed starting in the mid-1980s.

Although the two strategies operated simultaneously at the international and domestic levels, the economic nationalist strategy focused more of its adjustment efforts at the domestic level, whereas liberal continentalism emphasized the international level. The other key points of contrast between the two strategies had to do with the type of policy instruments they employed, the designated objectives of their plans of action, and the institutional framework that each industrial adjustment strategy required for its implementation.

The strategy of economic nationalism had three components. At the international level, the strategy promoted trade diversification in order to enhance trade with countries other than the United States and diversify the country's export commodity profile so that technology-intensive, highly processed commodities may represent a larger share of total exports. The second aspect related to asserting domestic control over Canada's economic activities (the state as gatekeeper). There were two key policies pertaining to this goal, Canadianization and the regulation of foreign investment. Canadianization refers to those government actions aimed at enhancing Canadian control and ownership of domestic industries. Its purpose was to capture the economic benefits associated with industrial development that would otherwise have been retained by foreign interests. Whereas Canadianization sought to reduce Canadian alienation from the process of capital accumulation, foreign investment regulation sought to channel the domestic activities of foreign investors so that they better served the industrial needs of Canada and, generally, to minimize the cost associated with the partial loss of domestic managerial control. The Foreign Investment Review Agency (FIRA) was created in 1973 to regulate the entry of foreign direct investment.

The last component of the strategy was the promotion of industrial development (the entrepreneurial state). In particular, Ottawa sought to expand the economy's industrial manufacturing and technological base in what was, up to then, a dual economy with a sizeable natural resource sector. To fulfill this task, the federal state created the Canada Development Corporation (CDC), using it to invest in high technology and process-intensive natural resource activities. The CDC was also used to undertake Canadianization, facilitating the transfer of corporate decision-making power from foreign interests to domestic interests via an active program of enhancing Canadian participation in the economy.

The strategy of liberal continentalism had two policy components. First, this strategy sought to institutionalize Canada-U.S. bilateral trade. The failure of the strategy of economic nationalism, increased vulnerability to American trade actions, and improvements in Canada's international position combined to make this policy option appealing after having been rejected for over a century. By expanding and securing market access, the 1988 Canada-U.S. Free Trade Agreement (FTA) and the 1994 North American Free Trade Agreement (NAFTA) would help bring both cross-border integration and specialization, which, in turn, would contribute to productivity gains, export growth, and high-skill job creation.

Politically, the free trade agreements' dispute settlement mechanisms were aimed at rendering U.S. trade policies more predictable, transparent, and amenable to some outside influence. There was another political dimension to the free trade arrangements. Free trade was an economic solution to the political impasse between the provinces and the federal state that was instigated in part by their interventionist tendencies in jurisdictionally-disputed public policy areas. Free trade facilitated the mutual withdrawal of provincial and central governments from some of their economic spheres.

The other component of liberal continentalism was to expand and sharpen market processes in the economy through privatization, deregulation, and elimination of inter-provincial barriers to trade. In sum, liberal continentalism was based on the strategic withdrawal of state involvement from the economy and the establishment of rules and regulations that institutionalized the operation of market governance both at the regional and domestic levels.

The Nature of Canada's Industrial Adjustment Problem

The industrial adjustment challenges that Canada confronted in the late 1960s and early 1970s had three aspects. First, tariff protection and the high concentration of American investment in Canada's secondary manufacturing and resource-based manufacturing created structural deficiencies in the country's economic system.[1] Second, Canada's economic position in the inter-

---

1. Secondary manufacturing includes such medium- and high-technology activities as chemicals; man-made fibers; industrial machinery; mechanical handling equipment; agricultural machinery; transportation equipment and vehicles; aircraft and parts; locomotives and rolling stock; communications equipment; heating, refrigeration, and air conditioning equipment; tools; office machinery; pharmaceutical supplies; photographic goods. Resource-based manufacturing includes activities such as wood products; paper products; primary

national economy deteriorated, as illustrated by the decline in its share of world manufacturing. As Britton and Gilmour (1978, 41) point out, "Compared with other countries Canada's improvement [in the 1950s and 1960s] was a weak response to a world economic environment that was remarkably conducive to economic growth and expanded trade." Canada's share of world manufacturing dropped from 6% in 1965 to 4% in 1975. Finally, Canada's increasingly close economic ties with the American economy became a cause of concern because it elevated the risk of becoming vulnerable to American policy actions and it produced differential levels of economic gains.

By the late 1960s, Canada's secondary manufacturing base displayed a number of features that gave strong indication that this sector was not performing well. The symptoms involved inefficient industrial structure, weak export capacity, heavy reliance on imported technology and manufactured goods, and the presence of a small indigenous managerial class. These symptoms came about, in large part, as a result of developing Canada's manufacturing base behind tariff walls and relying heavily on foreign direct investment. As instruments of industrial modernization, tariff protection and foreign investment helped develop Canada's manufacturing base, spur sustained growth that translated into high standards of living, and create competitive advantages in particular international markets. The tariff provision of the 1879 National Policy helped substitute domestic products for imported goods within Canada's expanding home market, which was facilitated by the National Policy's two other goals—promoting western settlement through an active immigration policy and creating a national railway system to connect the country's vast territory.

Paradoxically, the policy efforts intended to secure Canada's economic independence had the opposite effect. The lack of import competition led to the emergence of many inefficient firms in the manufacturing sector that were unfit to compete in international markets. Moreover, Canada's relatively small domestic market prevented many of these firms from achieving the necessary production runs to lower production costs. Import tariffs also spurred an influx of large amounts of American investment into Canada, which over time enabled American-owned Canadian affiliates to gain majority control of key industries and a solid presence in others. The import tariffs and Canada's promising future as an industrial power combined to create a powerful incentive for American investors to jump the tariff wall and establish branch-plants.

The extent of American participation in the Canadian economy by the late 1960s was unequaled by any other country that hosted American investment. Accordingly, if any country was in a position to fear that its economy was falling too deeply under foreign control, it was Canada. As Gilpin points out, "Countries and even all of Western Europe may worry about foreign

---

metal products; non-metallic mineral products; petroleum and coal products. These definitions are from Britton and Gilmour (1978, 198-9).

domination, and particularly American domination, but Canada is the only country where one can say with a considerable degree of truth that American corporations have taken over the economy" (Quoted in Britton and Gilmour 1978, 104). Table 1.1 shows that non-residents acquired majority control in the manufacturing, petroleum and natural gas, and mining and smelting industries. Although American investors fell short of securing majority control in manufacturing, American participation in this sector figured prominently if measured as a percentage of assets controlled by non-residents.

Table 1.1: Distribution of Ownership and Control of Canadian Industries, 1954 & 1967
(percentage)

Year	Canada		United States		Other	
	Owned	Controlled	Owned	Controlled	Owned	Controlled
Manufacturing						
1967	48	43	44	45	8	12
1954	53	49	37	41	10	10
Petroleum/Natural Gas						
1967	38	26	51	60	11	14
1954	40	31	57	67	3	2
Mining/Smelting						
1967	39	35	51	56	10	9
1954	44	49	48	49	8	2
Utilities						
1967	81	95	18	5	1	--
1954	86	92	12	7	2	1

Source: Government of Canada, *Foreign Direct Investment in Canada* (Ottawa: Information Canada, 1972), p. 20.
Notes: Utilities include air, road, water and urban transportation; telecommunications; and hydro-electricity. Manufacturing includes the following industries: food and beverage; tobacco; rubber products; leather products; textile; furniture; printing, publishing and allied; paper and allied products; primary metals; metal fabrication; machinery; transport equipment; electrical products; non-metallic mineral products; chemicals and chemical products; and miscellaneous manufacturing.

Aitken (1961) observes that while foreign investment did help Canada to achieve its developmental objectives, it also generated costs. Because of its sheer volume, American investment shaped the direction and pace of economic progress in Canada. Nowhere was this more evident than in the orientation that the expansion of the secondary and resource-based manufacturing assumed. Whereas American-owned Canadian subsidiaries in the secondary manufacturing sector employed their control to suppress exports, enhance intra-firm imports, and serve Canada's internal market, their managerial control in the natural resource industries was used expressly to serve the industrial needs of the United States. The inward-oriented characteristic of Canada's secondary manufacturing and the export-oriented feature of the natural resource sector were the consequence of the large presence of American investment in the Canadian economy.

While the export performance of the resource-based manufactured sector was strong from the 1950s to the mid-1970s, that of the secondary manufacturing sector was weak. Canada consistently maintained a trade surplus in resource-based manufacturing activities, with the United States serving as the key destination for commodities in this group. The trade balance in fully manufactured products experienced a growing deficit that reached a little over $10 billion in 1975, marking a three-fold increase from 1970. Moreover, exports of end-products made up only 32.7% of total exports in 1968, making Canada the weakest exporter in this category among industrialized countries.

This trade pattern was much influenced by the preference of foreign-owned subsidiaries to service the domestic market rather than export what they produced. Often, this practice was reinforced by establishing licensing arrangements that restricted Canadian subsidiaries from exporting the goods they produced. According to a governmental report, *Foreign Direct Investment in Canada* (Gray Report), 58% of American branch-plants and 43% of other foreign subsidiaries faced export restrictions imposed by their parent companies in the 1960s. In comparing the export practices of Canadian-owned and American-owned companies, Britton and Gilmour (1978, 110) note that "clearly foreign-controlled subsidiaries do not direct themselves to making export sales, whereas Canadian-controlled secondary manufacturing firms are the main ones penetrating foreign markets."

In addition to displaying export resistant tendencies, Canada's secondary manufacturing sector became highly dependent on imports of technology and other intermediary goods by the late 1960s and early 1970s. End-products constituted the largest group of imported goods and its volume registered the largest increase between 1968 and 1973 (Department of Foreign Affairs and International Trade 1996, 9). Moreover, imports in four high-technology industries—consumer electronics, computer and office equipment, general machinery, and high-technology manufactures—all listed very high levels of import dependence (Britton and Gilmour 1978, 45-9).

The high concentration of American investment in Canada's capital goods industries had a profound effect on these industries' import propensity. Concerning the 217 foreign-controlled companies operating in Canada in 1969, almost 50% of the materials they required were imported and only 30% of the goods they produced were exported (Britton and Gilmour 1978, 115-6). Even more significant was that intra-corporate sales ("tied imports") as a percentage of total imports in the secondary manufacturing sector was at 68.1%. This was also echoed in the Gray Report (1972, 183), which pointed out that "foreign-controlled companies tend to import from the country of the parent and indeed from parents and affiliated companies." It also remarked that "imports tend to be high in the sectors where foreign control is high [and that] foreign-controlled companies appear to be more import oriented than Canadian-controlled companies."

Perhaps the most significant effect this development had on Canada's industrial structure at the time was that it impeded the creation of upstream industries that could supply the intermediary products used in downstream production activities—that is, the end-products sector of the Canadian economy. The underdeveloped nature of upstream industries contributed to the warehouse-assembly processing characteristic of Canada's manufacturing sector in which basic assembly functions were carried out in Canada using imported higher value-added inputs. Moreover, for many of these branch-plants, the bulk of research and development activities were reserved for the parent company, thus impeding the development of an indigenous R&D capacity in Canada.

The presence of an entrepreneurial gap was another symptom of Canada's weak manufacturing sector in the late 1960s and early 1970s. The activities of foreign-controlled companies resulted in reducing the indigenous managerial class. The adverse effect of the entrepreneurial gap on Canada's secondary manufacturing was of some significance. By placing citizens of the parent company's home country into managerial positions in Canada, the indigenous managerial class lost out opportunities to develop its managerial and entrepreneurial capacity. The data indicate that the number of managers that were Canadian citizens increased as the share of foreign ownership of firms grossing more than $25 million decreased. But companies that were 50% or more owned by foreigners were less inclined to employ Canadian citizens as directors or presidents. In fact, only 44% of directors in those companies were Canadian citizens. The same pattern was repeated with respect to corporate presidents. Of the firms that were 50% or more foreign-owned, only 45% of presidents were Canadian citizens. In terms of retaining economic autonomy, these figures were alarming because the managerial decisions most closely linked to the pattern of industrial development in Canada were made by foreigners, whose preferences were more likely to be coincident with advancing the interests of their parent company and home economy than those of Canada.

Canada's sense of vulnerability to the policy actions of the United States increased starting in the 1960s. The problem of extraterritoriality—which arises when foreign companies obey and implement the laws and policies of their home governments in the host country—particularly concerned Ottawa. The Report of the Task Force on the Structure of Canadian Industry (Watkins Report) noted the following:

> The most serious cost to Canada of foreign ownership and control results from the tendency of the United States government to regard American-owned subsidiaries as subject to American law and policy with respect to American laws on freedom to export, . . . anti-trust law and policy, and . . . balance of payment policy. The intrusion of American law and policy into Canada by the medium of the Canadian subsidiary erodes Canadian sovereignty and diminishes Canadian independence (Watkins Report 1968, 360-1).

Between 1950 and 1971 there were twenty-four bilateral dispute cases that arose as a result of the United States government claiming jurisdiction over the behavior of American-owned multinational corporations operating in Canada (Leyton-Brown 1974). The most politicized and visible cases involved the extraterritorial issues of antitrust, export controls, and balance-of-payments policy. Of particular importance were the 1965 and 1967 balance-of-payments cases. The balance-of-payments form of extraterritoriality to which Canada was exposed resulted from American efforts to shift to its partners some of the cost of managing its growing balance-of-payments deficit. The United States' 1965 Voluntary Cooperation Program and the 1968 Mandatory Direct Investment Guidelines established specific goals to guide the domestic and foreign activities of American multinational corporations. The Voluntary Cooperation Program established voluntary guidelines, which called upon approximately 900 American multinational corporations to lower their net capital outflow by exporting more from the United States, to repatriate foreign earnings, and to raise investment capital abroad. The 1968 measure converted the 1965 voluntary guidelines into mandatory ones and they were to inform the activities of all American multinationals (Watkins Report 1969, 336-8; Wright and Molot 1972).

Although Canada was granted exemptions in both cases, they were granted with the understanding that Canada would take voluntary measures to assist the United States in rectifying its balance-of-payments problem. On the issue of repatriation of profits, the two governments came to an understanding that American subsidiaries operating in Canada were not expected to carry out significant repatriation of their earnings, an important amendment given that retained earnings were a major source of investment capital for Canada. Although Canada also obtained an exemption from the 1965 guideline that restricted American companies from exporting capital to foreign countries, it agreed to lower its reserve ceiling by $100 million as a way to lessen pressure on the American dollar. As for the 1968 guidelines, Canada obtained an unconditional exemption, but it voluntarily agreed to convert $1 billion of its foreign exchange reserves into American Treasury securities. Moreover, the United States managed to gain Canada's commitment to prevent banks, insurance companies, and other private investors from using Canada as a 'pass-through' to transfer American dollars to the Eurodollar market (Wright and Molot 1972; Leyton-Brown 1980-1).

The sharp increase in the United States' balance-of-payments deficit plus inflationary pressures compelled American President Richard Nixon in August 1971 to implement an aggressive economic strategy. It included instituting wage and price controls, which Chrysler and Douglass Aircraft Company enforced through their Canadian subsidiaries; imposing a 10% surcharge on imports into the United States, which affected the commercial activities of American-owned Canadian subsidiaries; and closing the gold window in an effort to stem the outflow of gold and restore some national policy autonomy. Unlike the two previous cases of extraterritoriality, Canada was not granted an exemption. The "Nixon shocks," as these measures came to be known, heightened Canada's sense of vulnerability to an unprecedented level.

The Nixon shocks were a deliberate effort by the Nixon administration to pressure advanced industrial countries to assume greater international responsibility in maintaining the international economic system. Moreover, it demonstrated that the United States was willing to employ aggressive unilateral measures to gain the acquiescence of its allies. For Canada, this did not bode well because it marked the end of the "special relationship" that previous leaders in both countries had evoked to help them steer through the complications of maintaining cooperative bilateral relations. The special relationship also had conferred upon Canada disproportionate commercial and security advantages that few other American allies were privileged to receive. In his address to the Canadian parliament in April 1972, Nixon called for the development of a "mature relationship," indicating Canada would have to increase its share of the cost of maintaining an open international economy and that both countries needed to recognize each other's right to pursue "autonomous independent policies" as well as "define the nature of [their] own interests." In short, a mature relationship meant that Canada could no longer rely on the benevolence of the United States.

The change in the politics of bilateral relations exposed Canada for the first time to the real costs of having a relationship of dependence with the United States. Then Secretary of State for External Affairs, Mitchell Sharp (1972, 7-8) remarked that "because of the high concentration of our trade with the United States...Canada was probably more exposed than any other country to the immediate impact of the U.S. measures and had more reason to be concerned about their future implications." He added: "They threw into sharp focus the problem of Canada's vulnerability."

The Nixon measures helped concentrate attention on the problem of industrial adjustment. By the early 1970s, Canada's industrial adjustment problem had two dimensions. First, the existing pattern of economic activities between Canada and the United States had shaped Canada's economy in such a way that it seriously undermined its capacity to pursue such economic goals as international competitiveness, technological progress, accumulation of wealth, and the development of an autonomous industrial base. The second dimension to the adjustment problem was that the prevailing pattern of economic relationship between Canada and the United States produced an unsustainable level of asymmetrical dependence. For Canada, the bilateral relationship was increasingly inhibiting its freedom of policy action and deepening its vulnerability to changes in American policies.

The Nixon shocks prompted Canada to rethink its adjustment strategy. From the early 1970s to the early 1980s, Ottawa tried trade diversification, foreign investment regulation, and an interventionist industrial policy. This path of policy development, however, produced limited positive results. In response to new opportunities and constraints in the mid-1980s, the Canadian government switched to a new policy path that sought to deregulate the domestic market and establish a continental institutional framework—with rules that would guarantee market access and facilitate the settlement of trade disputes. Since, industrial adjustment has been shaped by liberal continentalism. This book provides an account of why liberal continentalism emerged as the dominant policy framework.

Organization of the Book

The book is divided into three parts. In Part I, I develop a theoretical framework to account for the sequence of adjustment policy actions between the early 1970s and early 2000s, and provide a historical analysis of Canada's preexisting political economic institutions and of the role of the Canadian state in the economy. Chapter Two elaborates on the book's theoretical framework, which combines ideas from the realist theory of international relations and the historical institutional approach to path dependency. The last part of the chapter sketches Canada's preexisting political economic system, focusing in particular on organizational characteristics of the federal economic bureaucracy, federal system, and business. Chapter Three provides an overview of the evolution of the Canadian state's role in the economy by focusing of the three 'national policies.'

Part II focuses on the strategy of economic nationalism. While Chapter Four explores the limits of trade diversification, Chapter Five discusses the limits of state entrepreneurship and gatekeeping. In the two chapters, I will explain why Ottawa's efforts to carry out these three functions failed to produce favorable outcomes. Finally, Part III concentrates on the strategy of liberal continentalism. Chapter Six explains why Ottawa shifted to this strategy in the mid-1980s and why it was successfully implemented. Chapter Seven discusses how the strategy influenced Ottawa's adjustment actions and shaped the country's economic activities in the 1990s and early 2000s.

Chapter Two

Explaining Industrial Adjustment Policy

Various perspectives have been used to explain state action. The main challenge facing analysts studying this subject, as Moravcsik (1993, 9) observes, "is not *whether* to combine domestic and international explanations into a theory, but *how* best to do so" (italics original). The theoretical approach I develop in this chapter is informed by realist and historical institutionalist perspectives. Such a framework provides much leverage on explaining patterns of state behavior. In particular, the bi-level approach offers answers to two questions central to this book: what aspects of a state's industrial adjustment policy can be explained by what factors at what level, and why do the policy goals and intentions of a state change over time? This chapter is divided into three parts. Parts one and two elaborate on the theoretical framework; the last part provides an overview of the historical development of Canada's political-economic institutional framework.

Realist Perspective

Systemic approaches focus on those international forces that constrain and enable state actions. Although systemic approaches do not explain why states select one policy option over others, they do shed light on the pressures that the international system places on states and why states discount certain strategies while preferring others.

The realist imagery of the international system is one of continual positional competition among states, a consequence of the anarchical structure of the system. The interaction of states in a self-help system has enduring effects on the goals, preferences, and strategies of states. First, states are concerned about their security and independence and will seek domestic and international arrangements that help minimize threats to both goals.

Another concern that states have is the need to preserve or improve their relative position in the international system. When contemplating whether to take part in an international economic engagement, realism contends that states are particularly sensitive to relative rather than absolute gains.[1] State preference for relative gains stems from the recognition that if economic exchanges constantly bring disproportionate gains to certain participating states, they may in turn develop a dominating position over other participants. Grieco (1990, 46) observes that countries located in the medium power category, such as Canada, "may be extremely sensitive to gaps in gains, for they must simultaneously fear the strong and aspire to their status and they must worry that they might slip down into the ranks of the weak." Thus, to the extent that states seek to maintain their power position relative to other states, they will periodically reevaluate current economic interactions with other states to determine whether the distribution of gains is such that they are able to maintain their relative power position.

Another assumption that is central to the realist tradition is that state power influences the scope of state interests. As Rose (1998, 152) points out: "Over the long term the relative amount of material power resources countries possess will shape the magnitude and ambition . . . of their foreign policies: as their relative power rises states will seek more influence abroad, and as it falls their actions and ambitions will be scaled back accordingly." Thus, the diminution or augmentation of state power capabilities conditions what, when, and how state goals are to be pursued.

As briefly elaborated above, a realist approach to state action emphasizes the causal influences of state power and of the anarchical political environment in which states interact. Such an approach, however, is not without its shortcomings. The first drawback concerns realism's use of a worst-case/possibilistic logic to depict state behavior in the international system (Brooks 1998). According to this logic states are always concerned about the possibility of military aggression or domination by stronger states. As Mearsheimer (1994-95, 11) puts it:

> States in the international system fear each other. They regard each other with suspicion, and they worry that war might be in the offing. Although the level of fear varies across time and space, it can never be reduced to a trivial level. The basis of this fear is that in a world where states have the capability to offend

---

1. Waltz (1979, 105) observes that "when faced with the possibility of cooperating for mutual gains, states that feel insecure must ask how the gain will be divided. They are compelled to ask not 'Will both of us gain?' but 'Who will gain more?' If an expected gain is to be divided, say, in the ratio of two to one, one state may use its disproportionate gain to implement a policy intended to damage or destroy the other."

against each other, and might have the motive to do so, any state bent on survival must be at least suspicious of other states and reluctant to trust them.

It is expected that the fear of aggression and domination is more intense among militarily weaker states than militarily more powerful states. Because security is scarce in the international system, states will always assess their interaction with others in terms of what is optimal for improving their relative position. Accordingly, international cooperation rarely lasts over time, for the distributive effects of international engagements are bound to change unfavorably for some states and thus intensify their fear of military conquest or domination.

I argue that the possibilistic logic does not capture the interaction between Canada and the United States as well as it does in other bilateral relations in the world. States are animated more by a probabilistic logic than by a possibilistic one, as Brooks (1997) argues. The former logic is based on the premise that the international system generates more security than what structural realists argue, and consequently the fear factor that states face as a result of the condition of anarchy figures less prominently in the calculations of state officials. The probability of conflict is determined not just by the condition of anarchy but also by technology, geography, political regime type, and international economic conditions (Brooks 1997, 456). These factors help to determine the probability of military aggression by affecting the cost-benefit balance associated with the use of force by other states. Moreover, states pursue goals other than military security, such as political autonomy, economic independence, peace and justice, and domestic welfare. Accordingly, states will seek power and wealth precisely to advance these objectives, sometimes at the expense of military security.

Another weakness associated with structural realism is its overemphasis on relative gains. States are doubtless concerned about the relative achievement of gains, but it is questionable that states will always reject cooperative arrangements because the distribution of gains substantially benefits partners. The importance that states attach to either relative gains or absolute gains can vary over time; much depends on the circumstances that states are facing. For example, if the possibility of war between two states is absent, then absolute gains will take on a greater value than relative gains, but if one of the two states judges that it has become increasingly vulnerable to the actions of the other, the weaker state will be inclined to place more value on the goal of achieving relative gains.

The third shortcoming is that structural realism trivializes the importance of systemic processes, such as the intensity of interdependence. Non-structural variables or processes should be considered because, like distribution of power (structural), they create incentives and exert pressures on states and thus can change calculations of national interest (Nye 1988). For instance, interactions among states produce benefits, but they can also lead to undesirable effects, especially when states pursue divergent policies. As states integrate their economies, they are at greater risk of being adversely impacted by one another's uncoordinated policy actions. The more exposed states are to such type of policy action, the greater the incentive to do something about it, either by demanding greater coordination of policies or by reducing the degree of interdependence. However, some states may find it difficult to pursue the latter option because their relations with another state or group of states is so deeply rooted that there are no available alternatives to pursue at a reasonable cost. Thus, the states that are confronted with the condition of vulnerability interdependence will place a high premium on securing any arrangement that enhances the dependability of the relationship.

Vulnerability Dependence and Relative Power

In light of these shortcomings, I propose a modified version of realism that focuses on the degree of power asymmetry between a weaker state (Canada) and a stronger one (United States) and the intensity of interdependence. Since the 1960s, changes in Canada's position relative to the United States (systemic structural variable) and intensification of vulnerability interdependence (systemic non-structural variable) have altered the matrix of external opportunities and constraints that the Canadian state has faced. Below, I draw upon Yarbrough and Yarbrough's (1992) strategic organizational approach to elaborate on the effects of interdependence, and on Grieco's (1997 and 1996) "relative disparity shift" and "voice opportunity" hypotheses to explore the causal effect of relative power capabilities on state behavior.

Vulnerability interdependence arises from the closeness of inter-state relations and high costs that states face in changing the distributive terms of such relations. The strategic organizational approach contributes to the study of interdependence by contending that as more relation-specific assets are introduced in a trade relationship, the suppliers of such assets become increasingly vulnerable to the hazard of opportunism, which arises from the uncoordinated policies of trade partners. Relation-specific assets refer to capital investments that are for the most part used exclusively in a particular trade relationship. They include investments tailored to service a foreign market or participate in cross-border production processes, which can take the form of just-in-time supply chains or other arrangements where different stages of production are split up and located in different countries.

Relation-specific assets are what lead multiple industries in different countries to develop complementary, interlocking production structures with one another. These assets cannot be recovered or redeployed easily in the face of market disruptions arising from opportunistic behavior. Opportunism refers to those policy actions of a trading partner aimed at changing the terms of exchange in its favor without concern for the costs that such change imposes on other partners in the relationship.[2] The extent of country X's vulnerability to the opportunistic behavior of country Y depends on the degree to which X has invested relation-specific assets in its economic exchanges with Y, as well as on the level of country X's trade dependence on Y. If country X has allocated little relation-specific assets in its trade relationship with country Y and has a diversified trade profile, the latter's opportunistic behavior imposes limited costs on the former. On the contrary, if country X's industrial base is deeply integrated with that of country Y's and X's trade dependence on Y is high, Y's opportunistic behavior poses serious threats to X's economic security.

As Canada invested more relation-specific assets in the continental economic system and as its trade dependence with the United States intensified after the mid-1970s, so its sense of vulnerability to American unilateral economic policy actions increased, as did its desire to better manage the bilateral relationship. Blank (2005) vividly portrays how Canada's role in the North American economic system has evolved:

> Over the past 30 years, [Canada's] economy has been restructuring on a North-South axis. What characterizes the North American economic system and has differentiated it from Europe is the way in which industries have built integrated continental production, sourcing and distribution systems on regional specialization.

As Yarbrough and Yarbrough (1992) argue, because the hazard of opportunism increases as trading partners devote more relation-specific assets into a particular trade relationship, those partners face an increasing incentive to create a governance structure that has an effective third-party adjudication and enforcement framework. Such a regulatory framework restrains opportunistic behavior by any partner and provides for the redress of grievances. As Yarbrough and Yarbrough (1992, 35) note: "To guard against state-sponsored opportunism, states can enter the public-sector equivalent of private nonstandard contractual arrangements." Indeed, the central motive behind Canada's request for a free trade arrangement with the United States was to secure a set of legally binding obligations in which both sides signaled their commitments to providing enhanced market access to one another. A guarantee of market access would not only reduce the risk of further allocating relation-specific assets for Canada, but also render the strategy of integration an effective means of promoting industrial adjustment from now on.

Grieco's relative disparity shift and voice opportunity hypotheses help us identify the optimal time for a weaker state (Canada) to seek a collaborative

---

2. My definition of the term opportunism is consistent with Yarbrough and Yarbrough's (1992, 15) stating that it happens when "one country, by ignoring or cheating on its commitment to liberalization while other countries abide by theirs, may be able to gain at the expense of its trading partners."

economic arrangement with a powerful state (United States). The first part of this formulation is based on Grieco's relative disparity shift argument. This line of argument assumes that states are defensive positionalists, that is, states are concerned with how the distribution of gains—derived from a mutually beneficial arrangement among states—will affect their relative power position (Grieco 1990, 37-40). According to the relative disparity shift argument, a less powerful or secondary state will be less inclined to establish closer economic ties with a powerful state if the former has recently witnessed a decline in its relative capabilities in the context of a particular relationship. The prospect of closer economic ties tends to raise worries over whether the stronger state will increase its dominating position over the weaker state. In contrast, when the disparity in relative capabilities has been stable or has improved in favor of the weaker state in recent years, the fear of dominance by the secondary state will subside and become less of an obstacle to deepening bilateral ties with the powerful state (Grieco 1997, 175-77).

The second component of the power-based argument focuses on Grieco's "voice opportunities" hypothesis. The more instruments of power a secondary state can marshal the greater is its currency of diplomatic bargaining and the greater the probability it can secure those institutional terms that will allow it to exercise effective voice opportunities. Effective voice opportunities are, according to Grieco (1996, 288), "institutional characteristics whereby the views of partners (including relatively weaker partners) are not just expressed but reliably have a material impact" in a cooperative arrangement. Such an institutional capacity can allow a weaker partner to challenge the actions of a stronger partner using a built-in adjudication mechanism, to lock in distributional terms that are beneficial to it both politically and economically, and to influence and render more predictable the behavior of a stronger state through the arrangement's regulatory and coordinative mechanisms.

The Logics of Counterweight and Integration

As Figure 2.1 illustrates, bandwagon, strategic integration, domestic counterweight, and external counterweight are four types of state behavior, each one stemming from a unique set of international constraints and incentives. The degree of vulnerability dependence—which is directly associated with exposure to the hazard of opportunistic behavior—reflects both the level of Canada's trade dependence with the United States and of Canadian investments in relation-specific assets employed in the bilateral economic relationship. Trade dependence is measured by determining the share of total Canadian exports that go to and total Canadian imports that come from the United States. While it is difficult accurately to measure relation-specific investments, intra-firm trade and import content of Canadian exports are key indicators used in this study. Both measures capture changes in the level of cross-border industry integration as well as changes in the marginal cost of finding substitutes over time. This study contends that since the early 1970s, Canada witnessed a gradual increase in its vulnerability dependence with the United States—a consequence of rising levels

Figure 2.1: Classification of State Behavior

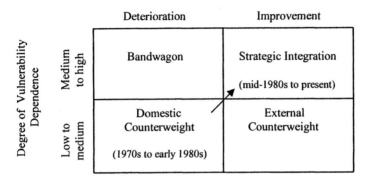

Relative Disparity Shift

of trade dependence and, equally important, of Canadian relation-specific investments.

Relative disparity shift refers to overall changes in Canada's power position relative to the United States. A general indicator of intra-regional shifts in relative disparities is cross-time changes in the Canadian share of the combined gross domestic products of the United States and Canada. Three other indicators are used to explore specific aspects of the unbalanced nature of the bilateral economic relationship. One of these is the extent of Canadian control over the productive sectors of its economy. Another indicator focuses on the Canadian presence in the American economy. One way to measure this variable is by assessing how diversified are Canadian exports bound for the United States and what are the value-added profiles of such exports.[3] The other way is by sizing up Canada's investment stakes in the United States.

The third indicator of intra-regional shifts in relative disparities focuses on policy conflict-induced changes in Canadian leverage relative to the United States. Of course, by having the world's largest economy, the United States has persistently wielded greater leverage over Canada than the other way around. But certain policy episodes in the 1970s and 1980s demonstrated that by asserting its political and economic independence, Canada could directly challenge American interests, as the Foreign Investment Review Agency and the National Energy Program (NEP) explicitly showed. As Doran (1982-83, 140)

---

3. For example, although approximately 30% of the American softwood lumber market in the 1990s was supplied by Canadian exports, in the event that this market was completely closed to Canadian exports it would amount to an annual loss no greater than a day and a half of trade. See Nossal (2001, 58).

points out, the policy measures linked to the strategy of economic nationalism "stand as reminders to the United States that . . . trade and commercial initiatives which ignore Canadian interests are likely to stimulate policies having very negative consequences for United States interests as well."

By drawing the U.S. government's attention, such policy conflicts can result in temporary gains in Canadian bargaining leverage, which in turn can be channeled to address bilateral issues of importance to Canada. Indeed, during the negotiations leading to the FTA, the structure of strategic bargaining was such that Canadian concessions on FIRA and NEP—whose costs were low in light of both programs' unpopularity—were reciprocated by American concessions regarding guarantee of market access, which Canada considered its primary objective.

These four indicators of relative disparity shift attempt to estimate changes in the unevenness of the bilateral economic relationship. In this study, I contend that Canada's relative position improved slightly in the 1970s and 1980s and remained stable subsequently, leading to a somewhat less lopsided relationship. If anything, the bilateral relationship has become more *inter*dependent since the early 1990s to the extent that Canadian intermediate and finished commodities and services, with high value-added profiles, play an integral part in the American economy and that Canadian investment is present in various sectors of this economy. Nevertheless, the relationship remains asymmetrical insofar as Canada's position in the context of the bilateral economic relationship is one that reflects vulnerability dependence, whereas the United States' position does not. That is, the costs of disruptions in bilateral economic flows are comparatively more damaging to the Canadian economy.

Mastanduno, Lake, and Ikenberry (1989, 468) argue that state power helps determine what adjustment arena (international or domestic) is selected. "Internationally weak states will emphasize domestic strategies more than will internationally powerful states. Likewise, powerful states will emphasize international strategies more than will weak states." Moreover, as the discussion of interdependence above suggests, the degree of vulnerability dependence that a state faces (in the context of a bilateral relationship) is likely to have an impact on how it responds to policy-induced disruptions in bilateral exchanges—that is, minimize liability by either deepening ties with third parties or developing bilateral coordinative mechanisms.

Two types of strategic behavior stem from the logic of integration: bandwagon and strategic integration. Bandwagon is based on a set of international constraints and incentives in which a weaker state enters into a bilateral agreement with a stronger state from a position of weakness, as depicted by a deteriorating relative position and above average level of vulnerability dependence. In contrast, strategic integration involves a weaker state entering into a bilateral arrangement under more favorable conditions, as illustrated by recent improvements in relative power position, but above average level of vulnerability dependence.

Two other types of strategic behavior are associated with the logic of counterweight: domestic and external counterweights. While both types of state behavior reflect below average levels of vulnerability dependence, the external type characterizes a state that leverages recent improvements in its power capabilities to "combat an overly close and dependent relationship with another country by intensifying the relationship with a third country or . . . with a regional bloc," as Dolan (1978, 28) points out. In contrast, domestic counterweight concerns a state that, because it lacks international leverage, attempts to develop internal equivalents of external counterweight, that is, develop domestically oriented adjustment policies that have the effect of lessening the influence of the more powerful state in its economy.

The strategy of economic nationalism was influenced by the logic of counterweight. During the 1970s, Canada's strategic behavior corresponded with internal counterweight, to a large extent. However, there is evidence that it was also informed by external counterweight, as the contractual links negotiated with the European Community and Japan in 1976 illustrate. Nevertheless, the strategy of economic nationalism was primarily geared toward domestic adjustment, leading to the development of various state interventionist roles— such as gatekeeping and entrepreneurship.

The change of adjustment strategy from economic nationalism to economic liberalism in the mid-1980s coincides with the slight improvement in Canada's economic position, at the same time that Canada's increasing vulnerability to American opportunism created a new set of systemic opportunities and constraints. This new strategic environment enhanced the political benefits of pursuing a strategy of integration, while increasing the costs of staying with domestic counterweight. Integration emerged as a worthwhile course of action for three reasons. First, it reflected the evolving nature of the bilateral economic relationship. Clearly, the "Third Option" failed to diversify Canada's trading pattern. Starting in the late-1970s, but especially since the 1980s, Canada began pouring more relation-specific assets into the bilateral relationship. Consequently, many Canadian and American firms were supplying each other with specifically tailored inputs, contributing to the development of highly complex North American supply chains.

Second, such a strategy could help Canada create an effective third-party enforcement framework to protect it against the gathering concern about American opportunism. The country's improved relative economic standing, it was believed, could be leveraged to lock in effective voice opportunities in a bilateral trade agreement—that is, create binding obligations, precise legal terms, and a dispute-settlement mechanism that satisfied Canadian interests. Finally, the strategy of economic nationalism was ineffective in achieving lasting economic adjustment. The selection of this strategy reflected the prevailing international constraints in the 1970s and domestic political support. Its failure was one of domestic implementation; the preexisting political economic institutions generated a political context not conducive to stabilizing the new policy regime. The next section elaborates on this last point.

Historical Institutionalism

Historical institutionalism has enriched our understanding of the study of politics by drawing attention to issues of timing and temporality in political processes. In doing so, this approach to institutions offers valuable insight into the mechanisms of social and political continuity and change. This study adopts Hall's (1986, 19) definition of institutions: institutions "refer to the formal rules, compliance procedures, and standard operating practices that structure the relationship between individuals in various units of the polity and economy." To say that "history matters" here is to claim that founding moments of institutional formation take countries along different developmental trajectories that are difficult to reverse. The next section elaborates on this claim by exploring the concept of path dependency. Then, in the following section I modify the theory of path dependency to provide sharper insights into the process of industrial adjustment policy changes in Canada.

Path Dependence

What sets path dependent process apart from other types of social processes is that it is self-reinforcing—that is, the cost of reversing an existing path of institutional or policy development increases with each movement along the path. Pierson's (2000a, 76) depiction of path dependent process involves three phases: first, a "critical juncture" in which events set in motion a movement toward a particular path out of at least two possibilities; second, the period of reproduction in which positive feedback mechanisms encourage continued movement along the path selected in the first phase; and third, the path comes to an end when new events unravel the current institutional equilibrium.

Pierson's description of historical political process—as one of innovation, reproduction, and breakdown—is what defines the punctuated equilibrium model of institutional change. Critical junctures or turning points—which can be triggered by economic downturns, partisan changes in government, or war, for example—create windows of time ripe for new policy departures or institutional changes. On the one hand, such moments create political space for policy entrepreneurs and the public to debate which of the various paths of institutional or policy development should be selected to cope with exogenous forces that disrupted the status quo. On the other hand, such moments generate forces for change strong enough to overcome the self-reinforcing dynamic that sustains existing policies (Hacker 2002, 59). Thus, critical junctures represent ruptures in long periods of relative stability that typically result in discontinuity or an off-path movement.

The outcomes of formative moments in the sequence become increasingly entrenched and reproduced over time. In Pierson's (2000a, 76) formulation of path dependence, positive feedback or increasing returns processes account for why a specific trajectory is difficult to reverse. Such self-reinforcing mechanisms, according to Pierson (2000b, 257) are not only "prevalent in politics," but are also "particularly intense" in this sphere. One feedback mechanism is that once individuals and organizations make large investments in a particular institution, they have an incentive to stick with it in order to recover those costs (high set-up or fixed costs). Another mechanism is learning effects; that is, repetition of tasks and functions improves individuals' knowledge and can lead to efficiency gains and further innovations. Third, the benefits an individual derives from engaging in a particular activity increase as other individuals align themselves with it (coordination effects). The fourth mechanism involves adaptive expectations effects; that is, when individuals expect others to adopt a particular option, they too adopt that option in order to receive the benefits accruing from such coordination.

Wood (2001) develops a causal account of why increasing returns processes are conducive to policy continuity. According to Wood (2001, 375), "Increasing returns processes engender a status quo bias *indirectly* by influencing the interests and preferences of actors in ways that incline them to support a particular institution or policy, and to mobilize in its defence if necessary" (italics original). The causal mechanism underpinning increasing returns consists of a two-stage process. First, institutions and policies producing learning, coordination, and adaptive expectation effects create incentives and constraints for individuals to act in certain ways. Second, these actors will seek to preserve the conditions conducive to positive feedback effects. This means that when policy entrepreneurs attempt to change policies, there will be overt political struggle between them and beneficiaries of existing policies, but also institutional resistance as expressed in the actors' reluctance to respond favorably to the policy entrepreneurs' call for their cooperation and commitment to new policies. Thus, policies guided by path dependence tend to encourage individuals to adapt their behavior to these policies in ways that push them further along the path of development.

However, this does not mean that policy change is impossible. Hacker (2002, 54) points out that the conditions favorable to positive feedback vary in intensity both across policy spheres and over time. Moreover, certain circumstances could undercut or overwhelm the self-reinforcing mechanisms that promote policy continuity. But the main claim of a path dependent argument is that bounded change tends to take place along an existing path of policy development. On this point, Pierson (2000a, 76) explains that "previously viable options may be foreclosed in the aftermath of a sustained period of positive feedback, and that cumulative commitments on the existing path will often make change difficult and will condition the form in which new branchings will occur."

A Modified Theory of Path Dependence

The theory of path dependence provides valuable insight into the analysis of industrial adjustment in Canada. A critical juncture or a turning point occurred in the late 1960s and early 1970s, triggered by deteriorating economic conditions. At that point, the policies that had guided economic adjustment since the early years of the postwar period (packaged in the so-called postwar Keynesian consensus) fell into disrepute, thus opening a policy window in which politicians and the public began to search for a new policy plan to promote industrial adjustment. The theory, however, does not provide much insight into what follows in the empirical story. Between the early 1970s and late 1980s, two distinct industrial adjustment strategies or paths of policy development were tried. Both represent off-path moves, for they exhibit 'logics' that are different from the one that prevailed prior to the occurrence of the critical juncture.[4] The strategy of economic nationalism was implemented in the early 1970s and abandoned in the mid-1980s, replaced by the strategy of liberal continentalism, which became entrenched and institutionalized. The theory of path dependence does not provide a clear answer to two related questions: What accounts for the sequence of policy moves? Why was the first path not stabilized by self-reinforcing mechanisms but the second one was? In answering these questions, I seek to modify the theory of path dependence to make it more applicable to this case study.

Path dependent arguments claim that in the early stages in a sequence, policy entrepreneurs and the public exercise relatively unfettered discretion in their policy selection. However, in various policy arenas, the latitude for action in the early stages can be quite restricted, offering actors few options. Thelen (1999, 385) argues that the model is "too contingent in that the initial choice (call it a 'critical juncture') is seen as rather open and capable of being 'tipped' by small events or chance circumstances, whereas in politics this kind of blank slate is a rarity. Not all options are equally viable at any given point in time." Similarly, Streeck and Thelen (2005, 8) observe that "there often is considerable continuity through and in spite of historical break points." One source of constraint is the international position of a state. In those policy spheres where the interplay of domestic and international forces assumes an important role, such as industrial adjustment, actors can tailor policy actions so that they are more or less internationally oriented. Powerful states enjoy more policy

---

4. I use the phrase "logic of a path" the same way Zysman (1994, 246-7) does; it refers to the "typical strategies, routine approaches to problems and shared-decision rules" that create predictable patterns of behavior by governmental and private-sector actors in a political economic system.

flexibility than less powerful ones insofar as they are able to determine the mix between a policy's inward and outward focus. Less powerful states, particularly those whose power capabilities have recently faded, have little choice but to focus on domestic measures. This systemic constraint, as I argued in the previous section, significantly narrowed Ottawa's choice sets in the early 1970s, leading it to select the strategy of economic nationalism.

Another factor that can constrain the latitude for action is the preexisting political-economic institutional framework. The institutions that this study focuses on are the organizations of business and the government, and the pattern of business-government relationship. When combined, they represent the institutional underpinnings of policies, or a policy regime.[5] This set of institutional structures, according to Zysman (1994, 271), "shapes the dynamics of the political economy and sets boundaries within which government policies and corporate strategies are chosen." It "induces particular kinds of corporate and government behavior by constraining and by laying out a logic to the market and policy-making process that is particular to that political economy." Especially, institutions shape the manner in which coordination—an integral component of industrial adjustment—occurs among private sector and governmental agents.

In liberal market economies, such as Canada's, coordination takes place via competitive markets. The private sector is characterized by arm's-length, contractual-based, and short-term relationships so as to minimize agency costs, maximize flexibility, and reduce market risks stemming from long-term obligations. Business-government relationship reinforces this market logic, and public policy tends to be biased in favor of market-based solutions. Overall, in this political-economic institutional system, actors "tend to be suspicious of collective support of economic action" and prefer "to liberate markets and contracts from social constraints and collective obligation" (Streeck 2002, 5 and 7). In coordinated market economies, such as Germany's and Sweden's, coordination occurs via strategic interaction. Both private sector and governmental actors seek to promote longer-term, obligation-based relationships. Both government-business relations and public policy are conducive to fostering networks of non-market coordination geared toward achieving collective obligations. The result is that this political-economic institutional system uses "various forms of hierarchical and organizational coordination that sometimes require heavy injection of public authority . . . often overriding contractual exchanges as entered into by private agents on their own volition, discretion, and calculation" (Streeck 2002, 6).

Different styles of coordination are necessary to carry out different industrial adjustment strategies. An adjustment strategy informed by economic nationalism—which seeks to redress the problem of industrial adjustment by subjecting economic decisions to greater state-led coordination—requires an institutional framework in which policy officials have the capacity to consult

---

5. This term is borrowed from Woods (2001, 256).

with and elicit consensus from actors whose cooperative responses are necessary to successfully achieve the adjustment goals of the state. On the other hand, economic liberalism—which seeks to address the problem of industrial adjustment by subjecting adjustment decisions to market governance—best performs in an institutional setting in which coordination is left to the private sector acting instrumentally within the parameters of market-enforcing regulations.

The viability of a new policy is determined, in part, by how well current institutional capacities match the tasks defined by that policy (Zysman 1994). If a policy's goal is to sharpen market mechanisms, the ideal institutional framework is one that facilitates coordination via competitive markets. If a policy's task is to strengthen networks of coordination, the appropriate institutional structure is one that facilitates coordination via strategic interaction. Where there is a mismatch or a suitable match between task and institutional capacity, the outcome is respectively an incoherent or a coherent policy regime.

The conditions that generate self-reinforcing mechanisms are weaker in incoherent policy regimes than in coherent ones because the institutionally-mediated behavior of actors fails to support the new policy's task. Coherent policy regimes are incentive compatible to the extent that the institutional context leads actors to independently adapt their behavior in order to reinforce the existing policy. The initial move toward the strategy of economic nationalism in the 1970s rendered the policy regime incoherent; but the subsequent move toward liberal continentalism restored its coherence. Consequently, the economic nationalist policy path stood little chance of becoming entrenched; but the prospect of stabilizing the path of liberal continentalism via positive feedback effects was much better. Still, this insight takes us only to a certain point in the explanatory account of Canadian industrial adjustment since the early 1970s.

Institutional accounts often run a risk of portraying institutions as highly constraining and immune to change. Structure trumps human agency; actors face limited degrees of freedom. While preexisting institutions do influence policy change, it is also the case that during critical junctures policy entrepreneurs will take advantage of the policy window to try to rework institutions so that they fit a new policy. Their ability to realign institutions to new policies—which is a necessary condition to trigger self-reinforcing dynamics—depends on their capacity to mobilize and maintain sufficient public support for their endeavor. As Deeg (2005, 174) points out, positive feedback effects do not spring up automatically when a new path of policy development is launched; instead, such effects need to be "cultivated." "Cultivation takes the form of mobilization in the policy arena on behalf of policy . . . change. It also takes the form of organizing collective action, often for the purpose of coalition building. Here power and ideas enter crucially into the institutional change process."[6] The goal of cultivating positive feedback effects, in this case, is to create new institutionally-mediated patterns of behavior that help stabilize the new policy regime.

The process of creating a new policy regime involves elements of "collective puzzlement" and "powering" (Hall 1993; Heclo 1974). In the initial stages in the path dependent sequence, when policy entrepreneurs wonder how effectively to address new collective problems, competing sets of policy ideas swirl around within the political discourse. These competing ideas put forth different interpretive frameworks that can help actors to understand the intricacies of the collective problems and direct them toward particular solution sets. The political discourse stays unsettled until a coalition of powerful state and societal actors is formed and endorses a particular set of policy ideas. At that point, the authoritative nature of the prevailing ideas is leveraged to adapt the institutional system in order to achieve the desired results (Hall 1993). Crucial to this process is that the new policy and the institutional modifications required to administer this new policy enjoy sufficient societal and state support to overcome the dynamics that reinforce the preexisting institutional structures, and that they become stabilized as these and other actors independently adapt their behavior to the new policy regime.

Between the early 1970s and mid-1980s, the political discourse was in a state of disarray. Inwood (2005, 7) describes the political discourse during this period as follows: "The post-war breakdown in consensus concerning Canadian economic development strategies had not been replaced by a new conventional wisdom. Rather, Canadian society was marked by sharp ideological polarization around the concepts of nationalism coupled with social democracy, and continentalism coupled with neoconservatism."

Collective puzzlement continued during the period in part because no set of policy solutions to Canada's problem of industrial adjustment mobilized a broad consensus; no coalition of interests was influential enough to be able to set the terms of political discourse and rise to a position of authority in the policy-making process. Because the coalition supporting economic nationalism was not as disorganized and weak as the one supporting liberal continentalism, the policymakers managed to push through institutional reforms aimed at upholding the state's role as gatekeeper and entrepreneur—the key policy instruments of the strategy. However, since the coalition lacked broad societal and state support, it could not fashion a coherent policy regime, that is mobilize a

---

6. Similarly, Gourevitch (1986, 20) notes: "Whatever [policy entrepreneurs] decide, their policies, to take effect, require compliance or even enthusiasm from countless individuals. . . . Politicians have to construct agreement from among officeholders, civil servants, party and interest group leaders, and economic actors in society."

sufficiently broad consensus to reinforce and entrench the new institutions recently created to administer the new strategy.

Starting in the early 1980s, the political discourse gradually gravitated toward a set of ideas advocating liberal continentalism. The disorganized, though growing, business-oriented coalition backing these policy ideas gained traction when the Macdonald Commission (see chapter 6) came out with a three volume report supporting free trade with the United States and a minimalist economic role for the state. The Commission's work helped strengthen the liberal continentalist-backing coalition by lending legitimacy to the policy ideas for which it stood. The fact that this alternative policy solution to Canada's economic problems enjoyed broad societal and state support and that its corresponding institutional reforms reinforced the type of coordination that the preexisting institutional structure was accustomed to, facilitated its stabilization via self-reinforcing effects and helped render coherent this new policy regime. The institutionalization of this policy regime has been exhibited since the early 1990s by the adoption of policy measures and of institutional reforms by governments of different partisan stripes that, in effect, have entrenched liberal continentalism.

Political-Economic Institutional Framework

For an industrial adjustment strategy to achieve its goals, it requires political support and appropriate institutional capacities. Both requisites do not come automatically when there is a switch to a new strategy; instead, policy entrepreneurs have to mobilize public support and rework preexisting institutions so that they provide actors with the right sort of instruments to achieve new objectives. The aim of the last part of this chapter is to provide a general overview of the historical development of the institutional framework within which policy entrepreneurs in the 1970s and 1980s maneuvered in an attempt to fashion a new policy regime. The analysis focuses on the evolution of the central bureaucratic structure, the federal system, and state-societal relationship.

Economic Bureaucracy

The federal economic bureaucracy has never acquired expertise in the area of indicative planning. This lack of planning capacity has been the product of the state's unwillingness to gather vital information from industries, forecast economic changes and respond proactively on the basis of such findings. Nowhere has this been more evident than in the area of the manufacturing industry. Before 1963, the year Prime Minister Lester Pearson established the Department of Industry, Canada was without a bureaucratic entity specifically in charge of advancing the country's manufacturing interests.

Historically, the bureaucratic apparatus has not been guided by an internally coherent and autonomous mode of thinking. From the end of the nineteenth century until the Second World War, although the authority of the executive grew relative to that of Parliament, policymaking was largely the preserve of the Cabinet and involved the bureaucracy but minimally. However, this pattern changed during the Second World War because increased public demand for a variety of public policies overwhelmed the limited capacity of the Cabinet. Consequently, the relationship between politicians and bureaucrats changed, ushering in the "age of mandarins," in which a small group of well experienced civil servants were called upon to generate policy ideas and define courses of action (Sharp 1981; Granatstein 1998). If ideas and information are sources of power, then the mandarins, who provided the intellectual guidance for governments from the Second World War to the late 1950s and early 1960s, wielded much political influence. Mitchell Sharp (1981, 43-4), a former civil servant and member of the group of mandarins, makes the following observation:

> When I was a civil servant, I think it is fair to say that individual Ministers and the Cabinet as a whole depended more upon the advice of senior civil servants than they do today and they did so deliberately. When a difficult problem arose, the customary response was to refer it . . . to a committee of senior public servants. There was also a period during the war and in the immediate post-war years when influential public servants . . . were active promoters of new ideas and approaches that they persuaded their Ministers and the Cabinet to adopt.

The age of the mandarins, however, was short-lived. Beginning with Prime Minister John G. Diefenbaker (1957-1963) and Prime Minister Lester B. Pearson (1963-1968), and even more so with the governments of Pierre Elliot Trudeau (1968-1979, 1980-1984), senior civil servants came under increasing criticism for wielding too much influence in the policymaking process. The influence of bureaucrats was markedly delimited as a result of the introduction of 'rational management'—a policymaking model that sought to centralize executive authority. First initiated by Trudeau, this model influenced later bureaucratic reorganization plans in the 1980s and 1990s (Aucoin 1986; Savoie 2000).

The bureaucratic structure that existed in the 1960s and early 1970s was not conducive to aggregating public power and executing policies. Since the end of the Second World War, bureaucratic entities had grown in size—as a result of the expansion of public responsibilities in a host of new policy domains—and were becoming increasingly differentiated and isolated from one another. Moreover, the proliferation of bureaucratic entities triggered turf battles between newcomers and veterans. This problem was quite evident in the policy domains related to the economy. The structure of the economic bureaucracy in the late 1960s and early 1970s was highly unstable as new bureaucratic organizations (such as the Departments of Industry—which later became Industry, Trade and

Commerce—and of Regional Economic Expansion) were established, and as hitherto less influential economic-related departments (such as the Departments of Agriculture, Consumer and Corporate Affairs, and External Affairs) gained more prominence. The result of such structural reorganization was that it worsened the existing problem of scattered authority and intensified bureaucratic conflicts between the traditional center of power, represented by the Ministry of Finance and of Trade and Commerce, and the new players (see Phidd and Doern, 1978; French, 1980; Jenkins, 1983; Atkinson and Coleman, 1989).

Rational management introduced an element of coordination to an otherwise decentralized, fragmented, and inefficient bureaucratic structure. Attention shifted to cabinet committees and central agencies. By the early 1980s, in comparison to the British and American governments, Canada had "the most institutionalized system of cabinet committees [and] the most highly differentiated central agencies support[ing] ministers' efforts toward collective decision-making," as Campbell (1983, 351) observes. Ministerial autonomy was replaced by cabinet-level decision-making, informed by the principle of collegiality and joint responsibility. By shifting decision-making from individual ministers to cabinet committees, it was expected that line department bureaucrats would not be able to influence individual ministers as much as they used to, and that this new institutional location would facilitate policy coordination, planning, and innovation (Aucoin 1986; French 1979). Bureaucratic influence at the agency and departmental levels was further curbed because the increasing number of bureaucrats working within the central agencies that served the cabinet and the prime minister created a counterbalance against line department bureaucrats (Aucoin 1986, 8).

Although rational management enhanced executive centralization, it did little to resolve the problem of fragmentation. There were two contributing factors. First, the problem of bureaucratic pluralism, which already existed at the departmental level, was reproduced at the central agency. For example, to the list of central agencies—the most important ones being the Privy Council Office, the Prime Minister's Office, the Treasury Board, and the Department of Finance—two new ones were added: the Ministry of State for Social Development and the Ministry of State for Economic and Regional Development. Second, too much confidence was placed on the norm of collegiality to facilitate coordination and collective decision-making. As past experiences illustrated, such achievements resulted from prime ministerial leadership. Lacking prime ministerial leadership, ministers were reluctant to adhere to the norm of collegiality and tended to revert back to pursuing the objectives of their own departments.

This sketch of the organizational structure of the government in the twentieth century offers two key insights about the general institutional dynamics of Canada's federal executive. First, fragmentation and lack of coordination were inherent characteristics of the federal executive. However, the periodic institutional reforms aimed at solving these two problems inadvertently created a new one—namely the relocation of political authority within the executive. The consequence of such reforms, as Atkinson and Coleman (1989, 64-5) point out, was that "no organizational home exist[ed] in the executive institutions of Canadian government . . . from which consistent political and bureaucratic instructions [could] be expected to flow." Ultimately, those institutional flaws undermined bureaucratic efficiency and effectiveness.

The other insight concerns the relationship between politicians and economic bureaucrats. With the rare exception of the "civil service mandarins," department-level bureaucrats did not enjoyed the kind of relative autonomy that facilitates long-term planning, fosters a common mode of collective thinking within the economic bureaucratic apparatus, and permits the exercise of statutory discretion with respect to the allocation of resources in society. Instead, constrained by the short-term political control of elected representatives, bureaucrats used their limited authority and capacity to serve their political masters.

The Federal System

The evolution of Canadian federalism has been anything but a smooth one. The path of development has taken Canada through pendulum-like swings between centralist and decentralist tendencies, between intergovernmental cooperation and competition, and between security and insecurity of jurisdictional authority. In contrast to unitary states, federal states divide political authority and responsibilities between constituent units and the central government and provide for the representation of regional interests in the central government. The purpose of federalism is to accommodate regional differences and enable regional interests to promote their values within their territorial space while simultaneously preserving the integrity of the nation-state.

Territorial differentiation, or regionalism, has been a permanent and prominent feature of the Canadian political landscape since Confederation. According to Smiley (1977, 450), regionalism has worked its effect on the political system by creating a "bias in favour of mobilizing interests which are territorially based and frustrating the political expression of interests and attitudes which are largely non-territorial." Regionalism is the product of ethnocultural differences between the French and English Canadians but also of economic geographical differences between the hinterlands and the center. Regionalism has been a source of tension in intergovernmental relations in part because of the difficulty in reconciling and accommodating regionally specific interests, and in part because of the spatial impact of federal policies. For example, the resource-based hinterland provinces have faulted the federal government for placing them in an economically disadvantaged position relative to Ontario and Quebec, which have served as Canada's industrial center (Bickerton 1999; Cairns 1977; Smiley 1977).

Unlike the German federal system, in which functional federalism is an important mode governing intergovernmental relations, the dominant mode in Canada is jurisdictional federalism.[7] Whereas functional federalism links the division of power to the logic of division of labor, jurisdictional federalism handles power in accordance to the logic of division of jurisdiction (Chandler 1987). In the German federal system, functional federalism is such that in several policy areas, the central government enacts laws and the Länder— Germany's constituent units—implement and administer laws (Painter 1991; Michelmann 1986). Federal systems modeled on this type of arrangement tend to have institutionalized coordinative mechanisms to ensure a smooth operation of the functional breakdown of the policymaking process between the two orders of government.

In Canada, although all residual powers belong to the federal government and thus contribute to the federal government's vast range of constitutional powers and responsibilities, the Canadian constitution also endows the provinces with powers and responsibilities that enable them to have a commanding role in such areas as provincial economic development and management of natural resources, as well as in the provision of welfare and education. Canadian federalism has induced the provinces and Ottawa to be as functionally similar as possible, and to maintain as much freedom of action and autonomy as possible within their own constitutional spheres of jurisdiction. However, in several public policy areas, most notably industrial adjustment, the constitution does not allocate exclusive responsibility to either level. Consequently, when the two levels of government are involved in overlapping jurisdictions, jurisdictional federalism has tended to heighten intergovernmental wrangling.

A brief overview of the evolution of Canadian federalism should suffice to demonstrate how the logic of jurisdictional federalism has played an integral part in the evolution of the federal structure. The period from 1883 to the 1910s witnessed the ascendancy of provincial power and adherence to the "watertight compartments" interpretation of the constitution (Cody 1977). Described as the period of "dual federalism," the provinces benefited from various pro-provincial rulings of the Judicial Committee of the British Privy Council—Canada's final court of appeal at the time, located in Britain. By broadly interpreting provincial powers, the court limited in essence the scope of the federal government's supremacy clause and residual powers (Archer, Gibbins, Knopf, and Pal 1999, 149). The dismal performance of the economy during this period did little to improve the public's judgment of the central government. The economic depression that lasted until the late 1890s, when the wheat boom arrived, diminished the public's confidence in the central government's capacity to manage the economy (Black and Cairns, 1966).

---

7. German federalism combines both functional and jurisdictional federalism. The Basic Law, Germany's constitution, gives certain powers to the Länder as well as to the central government.

However, between the 1910s and 1960s, the social turmoil brought about by the Great Depression, the pressures of participating in two world wars, and the development of the Canadian welfare state strengthened Ottawa's position vis-à-vis the provinces and centralized the federal system. Two other developments spurred the centralist trend in Canadian federalism. First, the Judicial Committee of the Privy Council was abolished and replaced in 1949 by the Canadian Supreme Court, whose decisions tended to be more in favor of centralism. The second factor that facilitated the centralist trend in Canadian federalism had to do with the fact that the provinces were becoming more dependent on federal transfers. Rising demand for public services coupled with a small revenue base forced the provinces to turn to the central government for financial help. This latter development, in particular, ushered in a period of "cooperative federalism" leading to tax-rental and tax-sharing arrangements between Ottawa and provincial governments and, ultimately, to greater federal involvement in provincial affairs.

The emergence of province-building in the 1960s, however, reversed this trend (Black and Cairns, 1966; Young, Faucher, and Blais, 1966). Quebec's Quiet Revolution in the 1960s attacked the province's established social conservative norms and asserted francophone control over the economy and social spheres. In order to promote the latter goal, Quebec authorities strived to bring about economic modernization and demanded greater freedom of action and autonomy from Ottawa over all policy arenas that fell under provincial jurisdiction. Province-building swept to other provinces, as illustrated by provincial assertion of independence and expansion of ambitions. The emergence of province-building coincided with the expansion of bureaucratic and planning capacities and the rise of the spending powers of provincial governments. Since the 1960s, provinces have sought to move into, occupy, and shape those public policy areas that are not exclusive to either level of government and to resist federal intrusions into provincial spheres of jurisdictional responsibilities. The forces of provincialism and nationalism have competed in shaping the Canadian society and economy. Like many policy spheres, Canada's industrial adjustment policies have been strongly influenced by this dynamic. One of the effects of this development is that intergovernmental cooperation has become more difficult to achieve, requiring extensive consultation and sustained trust-building efforts on both sides.

The Organizational Structure of Business

In his seminal historical comparative analysis of industrialized nations, Gerschenkron (1962) discerns a relationship between the timing of industrialization (early or late industrializing nations) and patterns of interaction among business, banks, and government. Early industrializers, such as Britain, undertook economic development with limited government intervention in the economy and with limited reliance on bank credits to help private entrepreneurs move into new industries. For late industrializers, such as Germany, the process of economic development required more state intervention and greater reliance on bank credits to facilitate the development of new industries. According to Gerschenkron, two economic variables explain these differences between early and late industrializers. First, whereas the leading industrial sector that early industrializers sought to develop, such as textiles, was less capital intensive, the primary industrial sector that late industrializers sought to develop, steel, was more capital intensive and its viability rested on achieving large-scale production. Thus, the costs of launching industrial development for late industrializers, in terms of capital requirements and scale of production, was relatively higher than the costs with which early industrializers had been confronted.

The second variable concerns the level of competition that industrializing nations faced in the initial stages of economic development. Whereas early industrializers were pioneers in the industrialization process, late industrializers had to go head-to-head with earlier industrializers. High startup costs and economic uncertainty put a high premium on coordinated efforts among the state, banks, and industrialists. The economic role of the state differed among late industrializers. State involvement was more direct when the banking system was weak and undeveloped, while the role of the state centered largely on infrastructural development when the financial system was strong and effective in supplying capital to industrialists. Although a late industrializer, Canada did not develop the bank-business relations nor imitate, for the most part, the developmental practices of states in other late industrializers. Like its American counterpart, an arm's-length pattern of interaction emerged between the Canadian business community and financial institutions, on one side, and the government on the other. The existence of capital markets at the time of Confederation and their continuing growth thereafter, preempted the development of a credit-based financial system (Zysman 1983, 69-75).

The financial system that existed at the time of Confederation was biased in favor of underwriting the expansion of merchant shipping, exchange of commodities, transportation, and utilities. When manufacturing activities began to rival commercial ones, the financial community regarded investing in manufacturing as being too risky. Consequently, Canadian banks kept their involvement in the development of manufacturing activities at a minimum and in turn enhanced the role of capital market actors in industrial development. As Atkinson and Coleman (1988, 39) point out, Canada's financial system was marked by the "unwillingness of banks and other financial intermediaries to undertake long-term and equity investment in industrial firms." Instead, "[t]he chartered banks devoted themselves primarily to short-term personal and corporate lending, while long-term capital [was] raised by industrial corporations in securities markets." The dominance of the capital-based financial system—and by implication, the relative autonomy of industrialists from banks and the state—eliminated a key channel through which financial institutions and the Canadian government could have influenced the allocation of investment capital for the purpose of achieving long-term strategic developmental goals.

The issue of trade protection figured prominently in the early years of Canadian industrialization. Until the early 1900s, business was polarized between the supporters of tariff protection—primarily manufacturers—and the partisans of free trade—namely, export-dependent firms involved in timber, dairy, and mining activities, and those dependent on imported manufactured items. While free trade proponents were organizationally weak as well as regionally and functionally differentiated, supporters of protectionism were regionally concentrated and functionally undifferentiated—which accounts for their ability to create in 1874 the Ontario Manufacturers' Association, which later on became the Canadian Manufacturers' Association (CMA).

In the 1878 federal election, the Conservative Party rode to victory with the backing of protectionists. Subsequently, free trade supporters backed off and thus eliminated at least one of the main sources of division within Canadian business. As Bliss (1974, 96) points out, by 1911 "business comment on the trade question was overwhelmingly weighted in favour of protection, so much so that there was no real debate among businessmen on the question. The National Policy [of which tariff protectionism was a key component] was national business policy."

Over the course of the twentieth century, Canadian business was splintered into three groups—the staples fraction, the branch-plant manufacturing fraction, and the corporate nationalist fraction (Atkinson and Coleman 1988, 48-52).[8] In the absence of a well-organized labor and/or farmer's movement, the three fractions had little incentive to try to reconcile their differences. Historically, the staples fraction—led by an alliance among finance, transportation, and resource capital interests—was the most dominant, with roots going back to pre-Confederation times.

Initially, its activities were predominantly linked to the sphere of circulation, but over time they spread into the sphere of production—as illustrated by its entry into resource-related manufacturing such as steel, pulp and paper, food and beverages, and construction. Clement (1977, 24) points out that activities in the sphere of circulation—transportation, utilities, and finance—tended to be controlled by Canadian-owned companies, whereas activities in the sphere of production—which includes production of new staples such as oil, minerals, pulp and wood—tended to be controlled by American-owned Canadian subsidiaries. In 1976, the Business Council on National Issues (BCNI) was launched by this business alliance in response to Prime Minister Trudeau's executive reforms, which, as Gillies (1981, 72) observes, made it

---

8. Instead of this classification, Clement (1977, 25) uses a different set of terms: dominant indigenous fraction, dominant comprador fraction, and middle-range indigenous fraction, respectively.

more difficult to access policymakers and "much less easy for the businessman to identify the players with power."

The branch-plant manufacturing fraction included Canadian subsidiaries of American firms, whose main goal was to occupy and service the domestic market. Represented by the Canadian Manufacturers' Association (CMA), this business group supported the continuation of an import-substitution strategy, first implemented by the First National Policy. However, in the early 1980s, the CMA, following the BCNI, announced its support of a free trade arrangement with the United States. The two fractions had a history of joining forces when their interests converged on salient policy issues. For example, they created the Canadian Home Market Association during the 1911 elections to compete against free trade supporters and the Liberal Party, and established the Canadian Reconstruction Association after the First World War to drive home the point that protectionism was in Canada's best interest. In 1987, the Canadian Alliance for Trade and Job Opportunities was created by corporations from both fractions in order to promote a "fair and workable" free trade agreement between Canada and the United States as well as a multilateral trade liberalization (Langille 1987, 69).

The nationalist coalition of businesses was composed of Canadian-owned firms that strongly supported policies that enhanced national control of productive resources and promoted the international activities of Canadian firms (Noisi 1985). The firms in this coalition came from an array of industries— energy, mining, telecommunications, transportation, real estate, and media (Atkinson and Coleman 1988, 49-50). This coalition fared well in the 1970s when its ideas were picked up by the Liberal Party and incorporated into the federal government's industrial adjustment strategy. However, the coalition's influence began fading away in the early 1980s as members abandoned the group and as liberal continentalism gained more societal and governmental support.

Conclusion

Institutions and politics shaped the process of economic adjustment policy change in Canada. The legitimacy of the policy regime that had guided industrial development since the late 1940s eroded in the late 1960s and early 1970s when it became clear that the current set of policies had failed to generate desirable political-economic outcomes. A new coherent policy regime would not emerge until the end of the 1980s, more than fifteen years after the postwar policy regime had been abandoned. The theory of path dependence under-explores the dynamics that shape the early stages in path dependent sequences and tends to undervalue the role of political power and ideas in such processes. In the first part of the chapter, I developed a set of arguments that seek to account for the changes in the orientation (reliance on domestic or international adjustment measures) and logic (counterweights or strategic integration) of economic adjustment policies. Changes in Canada's level of vulnerability dependence toward the United States and in its relative power position influenced the government's judgment concerning the orientation and logic that should guide industrial adjustment.

The implementation of policies, as I argue in the second part of the chapter, is shaped by particular aspects of domestic politics—namely, the interplay of structurally generated interests and preferences of actors and the politics of building support for new policy regimes. These domestic forces proved unpropitious for the implementation of the strategy of economic nationalism, as exhibited in the futile efforts to fashion a coherent policy regime around this strategy. This claim is revealed clearly in chapters four and five, which explore the three policy objectives that defined the strategy of economic nationalism—diversification, gatekeeping, and economic development. In contrast, as chapters six and seven illustrate, the interplay of institutions and politics proved conducive to the construction and institutionalization of a liberal continentalist-based policy regime.

Chapter Three

The Role of the Canadian State in the Economy

Scholars often study the evolution of the role of the Canadian state in the economy through the lenses of three "national policies" that cover a period of more than one hundred and thirty five years. Although all three national policies were created with the intent to promote nation-building—in particular, industry, society, and infrastructure building—each one has tended to reflect a general "conception of the state's relation to the market, the instruments available to the state, and the political and economic constraints on state action" (Eden and Molot 1993). This chapter explores each national policy, but focuses particularly on the third one. I argue that the "Third National Policy" has had within it both statist and liberal variants. The statist variant was slightly more influential throughout the 1970s and early 1980s, but since then the liberal variant has become the most prominent one—the one by which the current national policy is defined.

The First National Policy, 1879-1920s

"The time has arrived when we are to decide whether we will be simply hewers of wood and drawers of water [or will] make this a great and prosperous country, as we all desire and hope it will be."

The above statement was made by Finance Minister Samuel Tilley at the time when Sir John MacDonald's government enacted the National Policy tariffs in 1879, a key component of the First National Policy. Tilley and many others after him have contemplated what kind of role the Canadian state should play in the economy. Political economists, such as Zysman (1983), have identified different forms of state involvement in the economy. In the state-led model, which has been followed in Japan and France, the state guides the process of economic modernization by promoting the development of particular industries via the selective provision of industrial credit, subsidies, and other policy instruments. In the company-led model, found in the United States and Britain for example, the state plays a "nightwatchman" role, ensuring that market governance facilitates the maximization of market actors' self-interests. Lastly, in the tripartite-negotiated model, which small European states have followed, the state attempts to foster a policy environment conducive to business-labor coordination, which has tended to translate into investment-led growth coupled with egalitarian outcomes.

Today, the pattern of state involvement in Canada matches that of the company-led model. However, before the 1980s, none of the models would have captured satisfactorily the characteristics of the state's economic role. Historically, both the "firm-centered culture" and the "public enterprise culture" have shaped the state's role (Aitken 1964; Atkinson and Coleman 1988; Hardin 1974). The firm-centered culture defends market-based outcomes and is critical of state initiatives that are restrictive and intrusive on the free play of market forces. As Atkinson and Coleman (1988, 42) point out:

> For Canadian business, the principle of competitive markets is enshrined to the point that the internal decisions of firms are not considered legitimate targets of political action. Under these circumstances, the prospect of a concertative relationship, in which business and government collaborate in making longer-term investment decisions, is greeted with hostility.

Although the market logic is an integral feature of the relationship between the Canadian state and business, another equally significant, historical legacy has influenced the pattern of relations between business and the state. The public enterprise culture that Hardin and Aitken identify provides the clearest clue regarding the accepted task and scope of government intervention in the economy. Hardin and Aitken observe that throughout Canadian history, state interventionism was largely influenced by geopolitical concerns. As Aitken (1964, 7-8) observes:

> No one familiar with Canadian politics . . . can question that defense against the economic and political dominance of the United States is today among the overriding concerns of the Canadian federal government, no matter which of the two major parties is in power. Resistance to the threat of absorption by the United States made such assistance vital, and, indeed, thrust on the government a continuously active strategic role in directing Canadian economic development.

Central to this "defensive expansionist" thesis is the idea that government officials should harness domestic capabilities in order to strengthen the country's economic sectors, build an integrated domestic market, and foster a sense of collective identity.

Throughout Canadian history, state interventionism has coexisted with free market enterprise. The geopolitical task of the state did not menace market players; the exercise of state capacity in the economy helped create an environment in which business could expand freely, profit abundantly, and make corporate decisions on the basis of market signals. The arm's-length pattern of business-government relations enabled the former to maintain its autonomy and impelled the state to build capacities that were limited in scope, but were appropriate to fulfill its geopolitical task.

The First National Policy of 1878 was a product of an alliance between the federal government and business, represented by financial, industrial, and transportation interests. For the Canadian business community, the First National Policy's three-pronged policy program of providing tariff protection, building a national railway system, and recruiting immigrants was instrumental for developing Canada's industrial base, achieving capital accumulation, and facilitating the creation of a home market. Although the First National Policy was an explicit display of the Canadian state's strategic exercise of political power, the policies aimed to encourage business to lead the way in the country's process of industrialization. For the federal government, the First National Policy was an instrument of survival—a way for a young country overshadowed by a powerful neighbor to ensure its territorial integrity and political autonomy. In essence, the political logic incorporated in the First National Policy directed the state to work rapidly to construct a national political and economic union of hitherto loosely linked regions.

Public ownership, state credit, and tariffs were the centerpiece of the First National Policy. The state used the first two types of instruments to build a national railway system and a national electric power grid. The construction of the Canadian railroad system in the late 1800s and early 1900s was owed in large part to state subsidies. In 1876, after much debate about whether to go through Maine, the government completed the high-cost, roundabout construction of the Intercolonial Railway, joining Halifax with Quebec and Montreal. In seeking to link western and central Canada, both federal and provincial governments underwrote—through land-grants, cash subsidies, and bond guaranties—the construction of several transcontinental railroads built by such private companies as Canadian Pacific Railway (CPR), the Canadian Northern Railway, and the Grand Trunk Railway. These efforts doubled the railroad mileage, led to the discovery of new mineral deposits thanks to the excavation work that preceded railroad building, and created numerous all-Canadian routes.

The rapid expansion of the Canadian railroad system, however, created a problem of overcapacity and consequently threatened the survival of many of these high-cost operations. Faced with the excess capacity and the precarious financial situations of many of these companies, the government set up, between 1918 and 1923, the Canadian National Railways (CNR) and brought under its ownership and control existing railroad companies, except CPR whose financial state was sounder.

Tariff protection became a permanent feature of Canadian economic policy until the 1970s. In defense of the National Policy tariff, Prime Minister John A. Macdonald observed, "There are national considerations . . . that rise far higher than the mere accumulation of wealth, than the mere question of trade advantage, there is prestige, national status, national dominion and no great nation has ever risen whose policy was free trade" (quoted in Macdonald 2000, 51). Ironically, while tariff protection helped the country to develop its industrial infrastructure, it did not help it to secure its economic independence from the United States. In fact, it encouraged American capital to avoid the tariffs by setting up operations in Canada. As Bliss (1970, 32) points out: "From the perspective of the late 1960s [the First National Policy] now appears to have been a peculiarly self-defeating kind of economic nationalism. The funny thing about our tariff was that we always wanted the enemy to jump over them. Some Walls!" By the early 1920s American investment exceeded the level of British investment in Canada, and by the mid-1950s, direct investment surpassed portfolio investment as the main form of American investment flowing into Canada.

Another factor that helped attract American capital was the fact that the Canadian economy in the late 1800s and early 1900s performed well. Canada's manufacturing base compared favorably with those of other "late-follower" industrializing countries such as Sweden, Italy, the Netherlands, and Japan. Throughout the nineteenth century, the quality of Canadian iron matched those of the United States, Great Britain, and Sweden (Laxer 1989, 11). Agricultural implements—such as harvesters, mowers, and ploughs—competed effectively in the markets of England, Germany, and France. Estimates show that in 1870, Canada was the eighth largest industrial country in the world, and in seventh position in 1900. Within the group of late-follower industrializing countries, Canadian exports of finished goods in 1899 totaled US$12 million compared to Sweden's $13 million, Italy's $23 million, and Japan's $10 million (Laxer 1989, 46). Moreover, in 1901, 33% of Canada's labor force was engaged in industrial activities compared to Italy's 35%, Sweden's 28%, the Netherlands' 36%, and Japan's 20% (Kuznets 1966, 106-7).

The presence of American capital in Canada during this period did not provoke the sort of anxiety seen in the 1960s and 1970s. Although nationalistic in their dispositions, neither the government nor the indigenous manufacturing class were too concerned with the extent to which American investors were gaining influence over the form and rate of industrial development in Canada. Canadian manufacturers focused almost exclusively on "occupying the home market" rather than on pursuing export opportunities and reducing their dependence on American technology and capital (Williams 1993). The true colors of Canadian nationalism remained up to the 1960s a peculiar mix, for in seeking to secure its political and economic survival in the shadow of an increasingly powerful neighbor, it invited this neighbor's industrialists to take part in

Canada's process of industrialization, thus giving them a sizeable slice of Canada's growing economic wealth.

The Second National Policy, 1920s-1960s

By the 1920s, the Canadian state had demonstrated a willingness to use public enterprises to produce desired changes in the economy. Such state activism, however, had not extended into the regulatory domain. This changed under the Second National Policy as a result of the expansion of the state's administrative capacities. Moreover, under the Second National Policy, the state added new instruments to its toolkit—such as tax-incentive programs and government-sponsored research and development—and began moving toward trade liberalization. The Second National Policy focused on a trilogy of challenges facing Canada: the cultural and economic implications of the growing economic influence of the United States in Canada, the implications of Canada's involvement in the two world wars, and the need to promote macroeconomic stability in the postwar period. Of the three challenges, the last one attracted lasting attention and accounts to a large extent for the changes in the form of state intervention during this period.

It was the logic of "defensive expansionism" that moved the Canadian government in 1932 to create the Canadian Radio Broadcasting Commission—later named the Canadian Broadcasting Corporation (CBC)—and in 1937 Trans-Canada Airlines—renamed Air Canada in 1964. According to the Privy Council Office (1977, 11), the creation of the Canadian Broadcasting Corporation was needed in order to prevent Canadian broadcasting from becoming dominated by broadcasts originating from the United States. Similarly, the reason for creating a national airline was to prevent "American airlines, which were already carrying Canadian travelers and airmail[, from becoming] the effective trans-continental carriers for Canadians" (Knubley 1987, 60).

However, coupled with these measures were a number of "defensive pragmatist" measures, representing a new brand of intervention (Knubley 1987). In response to periodic market changes in the wheat sector, the Bennett government created the Canadian Wheat Board in 1935, which was given a mandate to purchase and sell grains. The government also created a central banking system, headed by the Bank of Canada. Lastly, in 1936, the National Harbours Board was created to manage the largest harbors in Canada. By 1939, fifteen state-owned enterprises (also called Crown corporations) existed and were involved in such activities as transportation, communications, harbor management, and commodity marketing. The high demand for war supplies and munitions during the Second World War forced the government to turn to public enterprises. From 1939 to 1945, thirty-three Crown corporations were established specifically to assist in the war effort (Privy Council Office 1977, 12). Importantly, these wartime Crown corporations operated like private companies and were often headed by people with private sector experience. This model of public corporation was adopted by many postwar Crown corporations.

After the war, the government launched a plan to convert the wartime economy back to peacetime production, while maintaining industrial growth and preventing a drop in the employment level (Wolfe 1978). To facilitate peacetime conversion and promote macroeconomic stability, the government offered accelerated depreciation rates to private firms that purchased public owned plants, a transaction further made attractive given that such facilities were sold at reduced prices. The combination of tax incentives and discounted price appealed to American investors, who ended up purchasing many government-owned assets (Laux and Molot 1988, 53). These policy instruments were also employed to promote resource exploration and development. Again, American investors proved to be more taken by these incentives than Canadians, which explains why foreign ownership of Canadian resources rose steadily starting in the late 1940s.

In addition to dealing with the economic conversion challenge, the government had to come to grips with a growing balance-of-payments deficit. The deficit was produced by the rapid increase in imports of American capital goods. The root cause of Canada's balance-of-payments problem was the warehouse-assembly processing character of Canada's manufacturing sector, whose growth during and immediately after the war was contingent on continued imports of machinery, heavy equipment, and other inputs from the United States. Consequently, Canada became the world's leading importer of equipment, machinery, and tools in 1948 (Williams 1994, 114).

Canada dealt with its balance-of-payments problem by pursuing a two-sided strategy. The Emergency Exchange Conservation Act of 1947 imposed import controls, which established restrictions and quotas on imports of consumer goods, as well as demanding the licensing of imported capital goods (Wolfe 1978). Prior to the Liberal government's decision to take this measure, Alex Skelton, principal adviser to C.D. Howe—who was an influential cabinet minister between 1940 and 1957—had studied Canada's balance-of-payment problems and concluded that the government should change its industrial development strategy. In particular, he suggested that branch-plants be pressured to carry out more research and development activities in Canada, rely more heavily on Canadian producers to supply needed production inputs, and increase export of manufactured goods to the United States. Skelton's advice, however, was rejected by the government.

The import control policy, however, triggered a vicious circle. The policy ended up augmenting American presence in the Canadian economy, for it encouraged American firms to open up new Canadian-based branch-plants to manufacture consumer goods they previously exported to Canada. Because Canada's relatively small capital goods market could not meet the growing demand of branch-plants for parts, equipment, and other components, branch-plants were forced to import even greater amounts of such goods from the United States, thus worsening the deficit in the merchandise account.

The government also pushed for multilateral trade liberalization as a way to deal with Canada's balance-of-payments problems. In the 1947 trade negotiations of the General Agreement on Tariffs and Trade, Canadian negotiators fought hard for tariff reductions and obtained an agreement with the American government that led to the lowering of quotas on Canadian resource goods entering the United States. For C.D. Howe, it was worthwhile for Canada to secure market access to the United States in the area of natural resources because, as he saw it, "as the mineral reserves of the United States [were] diminishing, it was logical that attention [would] turn increasingly to Canadian mineral and associated hydro power resources" (quoted in Wolfe 1984, 60). Canadian energy and mineral resources was the focus of the 1952 Paley Report in which the Truman administration identified Canada as an important supplier of ten out of the twenty-two "key materials" that the United States needed to stockpile in order to ensure the country's national security and to sustain its industrial progress.

In 1939, Americans invested $4,151 million in Canada, of which $2,270 million (54.6%) was in the form of portfolio investment. By 1955, the figure reached a total of $10,295 million of which $6,513 million (63.2%) was in direct investment. American firms acquired major stakes in Canada's consumer goods industry as well as in various areas of the energy and mineral industry—such as lead, zinc, asbestos, iron ore, petroleum, nickel, and copper production. For example, in 1957, Americans owned a 75% share of Canada's mining and smelting production compared to 38% in 1946. Further spurring this inexorable continental pull was the completion of a major infrastructure project, the St. Lawrence Seaway, which connected the St. Lawrence River to the Great Lakes. Once a major waterway enabling Canada to trade with Europe, by the late 1950s the St. Lawrence River was becoming an important artery, which among its many uses, served to transport Labrador's iron ore to both Canadian and American steel mills of the Great Lakes region (Clark-Jones 1987).

In retrospect, the Second National Policy's lasting legacy was that it strengthened the north-south link by encouraging more American presence in Canada and deepened the east-west link by encouraging regionally specific forms of development. Importantly, the influx of American capital reinforced preexisting regional development pattern—with resource and mineral related investments concentrated in the western provinces and branch-plant manufacturing in Ontario and Quebec. This pattern of regional specialization, however, marginalized the Atlantic provinces. In sum, regional specialization was no longer a mere result of historical accident, it was purposefully promoted by both Canadian and American capital.

The Third National Policy:
Statist and Liberal Influences

The economic conditions in Canada by the late 1960s were mixed. On the one hand, Canada enjoyed a high standard of living thanks to a strong postwar economy, and developed a respectable international reputation founded on an outstanding peacekeeping record and contribution to the Western alliance. On the other hand, there was a growing economic malaise caused by the large presence of American capital in Canada, an export profile that was concentrated on the commerce of new staples, and a reliance on imports of capital goods, technologies, and managerial know-how. The Third National Policy responded to two general concerns. First, it offered an answer to the concern about the fact that if the current pattern of economic exchanges with the United States was left unchanged, Canada risked becoming increasingly vulnerable to American pressures. It became more and more apparent that the existing terms of economic exchange between Canada and the United States had contributed to the erosion of Canada's relative capabilities and margin of autonomy. The second concern to which the Third National Policy responded had to do with the need to increase the level of diversification and sophistication of Canada's industrial structure.

The political economic concerns that the Third National Policy addressed had been raised in the final report of the Royal Commission on Canada's Economic Prospects that came out in 1957. Walter Gordon, the chairman of the commission, observed that its purpose was to explore "the question of selling control of our business enterprises to foreigners and the effect this could have on Canada's independence" (quoted in Azzi 1999, 37). The Report acknowledged the benefits that American investments had made to the economic development of Canada. But it was also cautious, pointing out that Canadian interests were not necessarily advanced by or coincident with foreign interests. Canadians would not cease to be concerned about the large presence of foreign capital, the Report said, "unless something is done to make Canadian voices more strongly and effectively heard in some vitally important sectors of our economy in which non-residents exercise a large measure of control" (Gordon Report 1957, 399).

Although the Gordon Report received mixed reactions from government officials, the press, and professional economists, it had an important impact on the economic policy discourse, particularly in the late 1960s and 1970s. The Report helped sound off an alarm about the possibility that Canada's economic prospects could be jeopardized if the country continued to rely heavily on foreign investment and if industrial sectors continued to be dominated by branch-plant manufacturing, which tended to suppress innovation and technological advancement. Moreover, the Gordon Commission challenged postwar economic policy practices by introducing a rival set of economic ideas, interventionist nationalism, which called for a level of state involvement in the economy that went much further than what had been seen in the Second National Policy. Rather than the occasional macroeconomic fine-tuning, interventionist nationalism proposed that the federal state help increase indigenous involvement in the development of the manufacturing sector. In effect, the Commission helped define the main points of contention in the national debate that began in the late 1960s and early 1970s on the issue of industrial adjustment.

The Third National Policy marked a significant departure from the two previous national policies. From the late 1880s to the 1960s, four ideas guided the beliefs and policy actions of government officials with respect to state involvement in the economy and Canada's relations with the United States. First, Canada needed foreign investment if it was to exploit the maximum potential from its national resources. Second, the overriding concerns of the federal state were to promote national growth, improve domestic welfare, and secure national integration. Third, Canada's technological needs could be met by relying on technology imports at a minimum cost to the country's economic autonomy. Lastly, Canada's economic strategy was guided by the belief that neither Canada's export opportunities nor economic progress was undermined by the prevailing practice of branch-plant manufacturing in the economy.

By the beginning of the 1970s a new set of ideas had replaced the old one. Informed by the findings of several recent government reports, the Third National Policy advanced the following set of ideas. First, although Canada benefited from foreign investment, foreign investors needed to be held to stricter standards with respect to what they contributed to Canada's process of industrial development. At no point since the early 1970s did the Third National Policy seek to thwart or restrict foreign investment from entering Canada. Instead, the goal was to encourage foreign investors to be better partners with domestic firms in strengthening the economy's industrial capacity. Second, the state was to be more concerned about the relative achievement of economic gains; in particular, the distribution of those economic activities that enabled Canada to develop a sophisticated and diversified industrial structure, maintain and improve domestic welfare, and strengthen its national autonomy and political independence. This last concern was echoed by the Secretary of State for External Affairs, Mitchell Sharp, when he remarked that "there is no reason why [Canada] should not aim, in the context of an expanding economy and expanding trade prospects, to achieve relative shifts that, over time, could make a difference in reducing Canada's dependence . . . and, by extension, the vulnerability of Canada's economy" (Sharp 1972, 23).

Another idea guiding the Third National Policy was that Canada needed to foster an indigenous technological base to reduce its reliance on imports of capital goods. As the Science Council of Canada (1976-7, 26) urged, Canada needed to achieve "technological sovereignty," which it defined as the "ability to develop and control the technological capability necessary to ensure its economic, and hence its political, self-determination." The last idea held that

Canada needed to better manage the bilateral relationship in order to protect its national interest and, especially, to lessen its vulnerability to shifts in American policies.

The Third National Policy was not pursued in a uniform manner. On the contrary, I identify two variants associated with it: statist and liberal. Whereas the statist idea called for the state to play a greater interventionist role in the economy, the liberal variant called for the state to expand market governance. Below I elaborate on how these two influences diverged from each other by focusing on the differences in their policy orientation (see Table 3.1) and institutional requisites (see Table 3.2).

The Statist Variant

Under the statist variant, which came to embody the strategy of economic nationalism, the nature and scope of state involvement in the economy differed from previous decades. The limited opportunities available for problem solving at the international level reflected Canada's relatively weak international position. One of the underlying goals of this variant was to harness domestic capabilities in order to improve Canada's relative position. Indeed, Pentland (1982, 142) remarks that "the options paper recognized the organic linkage between the domestic and external dimensions of economic policy [but the] linkage was expected . . . to transmit the positive effects of domestic change outward to foreign policy—especially relations with the United States." Accordingly, the statist influence in the Third National Policy focused especially on expanding the entrepreneurial, gatekeeping, and economic planning roles of the state. The international undertakings associated with this strand were intended to lessen Canada's reliance on the American market by developing strategic counterweights to help diversify Canada's overall export profile.

The statist influenced adjustment policies required an institutional framework conducive to centralized economic policymaking and strategic coordination between business and government—which together would facilitate the implementation of medium- to long-term strategic plans. Shonfield (1970, 316) aptly indicates what institutional requirements are necessary to make such a strategy successful:

> Planning in a capitalist context . . . is a matter of tightening the hierarchical structure of government, compelling all departments to put all the decisions which have significant long-term consequences into a single intellectual framework, determined at the highest level of administration. New lines of authority are established, and at each level of power there is a more precise definition of the area in which choice and local initiative are allowed. Planning thus requires a high degree of explicitness in the relations between the different departments of government and a clear division of responsibilities.

The political-administrative organization that the statist variant of the Third National Policy required was found in postwar France and Japan until the 1980s. France's Planning Commission and Japan's Ministry of International Trade and Industry—the two central economic planning bureaucracies that were credited for orchestrating their country's postwar economic progress—evolved into internally cohesive and competent bureaucratic entities, and were given nearly exclusive mandates to plan and guide their respective countries' process of industrial development. In describing the role of the Canadian state, Sharp (1972) observed that it needed to be "deliberate," "involve some degree of planning, indicative or otherwise," and involve "at least a modicum of consistency in applying it." In further elaborating on the required domestic capacity of the federal state, Sharp added: "One implication of the conception of deliberateness is that the strategy may have to entail a somewhat greater measure of government involvement than has been the case in the past."

The state's relationship to society is another important institutional aspect of the statist variant. Once again, France and Japan are instructive cases to consider. The development of state-societal networks in France and Japan has traditionally brought state officials in close contact with particular groups of firms, and through the practice of "administrative guidance" in the case of Japan or "indicative planning" in the case of France, state officials have been able to gain the acquiescence of the private sector—by offering incentives and disincentives—for the purpose of coordinating industrial change (Shonfield 1970; Weiss 1998). This coordinative capacity of the state, as Weiss (1980) observes, is a crucial component of state capacity and is instrumental in producing industrial adjustment. More important still, the attention paid to a state's coordinative capacity signifies that state strength is not solely based on insularity from particularistic interests, but is also derived from its ability to elicit the consensus of the private sector.

For the statist variant of the Third National Policy to be effective in a federal state, an institutionalized pattern of intergovernmental cooperation, modeled on functional federalism, was ideal. Simeon (1979, 22-3) informs us about the difficulty in devising a coherent plan of action when intergovernmental relations are lacking coordination and division of labor:

> We have a clash at two levels. At the level of the substance of industrial strate-gy, the questions are: Where development will take place; what sectors will be emphasized; and how the conflict between the desire for regional growth can be reconciled with the desire to promote redistribution and maximization of aggregate growth across the whole country. At the level of procedures there is the question of how such a reconciliation might be brought about, and who is to make industrial policy. Is it Ottawa alone, the provinces acting independently, or some combination?

Table 3.1: Differences in the Policy Specifications of the Statist and Liberal Variants

Components of Policy	Statist variant	Liberal variant
Reduce and manage Canada's vulnerability dependence resulting from heavy reliance on American market	Diversification of export goods and markets	Establish a rules-based institutional framework to secure market access and manage trade disputes
Method of channeling dynamic forces in Canada's direction	Use of state interventionism	Use of market governance
Primary problem-solving arena	Focus on domestic means to economic adjustment	Achieve adjustment via international measures

Table 3.2: Institutional Characteristics of the Statist and Liberal Variants of the Third National Policy

Institutional arrangements	Statist	Liberal
Degree of centralization of policymaking process	Centralized policymaking with long-term, strategic planning capacity	Less centralized policymaking with little planning capacity
Type of federal structure	Functional and cooperative federalism	Competitive and market-preserving federalism
Pattern of business-government relationship	Strategic coordination aimed at promoting collective goals	Arm's-length mode of interaction aimed at enhancing free market enterprise
Types of state capacities	Capacity to strategically allocate resources to generate desirable industrial change; emphasis on capacity to mobilize consent	Strategic extrication of the state from the economy and imposition of market governance; centrality of deregulation, liberalization, and market allocation

Under functional federalism, the central government would formulate and decide on a national strategy of industrial adjustment, while the implementation and administration of the strategy would take place in coordination with the provinces. Moreover, to facilitate consensus building between the two levels of government on what goals a national adjustment strategy should pursue, the practice of cooperative federalism—widely used in the post-1945 period up to the late 1960s—would be imperative. Ideally, such an institutional forum would involve a small body of civil servants and political executives from each province and central government, and provide the context in which they act upon the recognition of opportunities for the attainment of mutual interests. Such a structure would dispose officials from the orders of government to think in terms of what is in the best interest of the nation and what the economic purpose of the state should be, and subsequently to undertake the task of drawing up a national strategy to improve the economic capabilities of the sum of the constituent units, rather than sub-national economies.

This overview of the institutional requirements of the statist variant gives us an idea of the extent and scope of the institutional innovations that the federal state had to undertake to make the strategy of economic nationalism work in an ideal way. At the end of the 1960s, few of these institutional features existed. In effect, the challenge in pursuing this adjustment strategy came in the form of institution building.

The Liberal Variant

The liberal variant of the Third National Policy or liberal continentalism tackled the industrial adjustment problems by expanding market mechanisms. Accordingly, the state interventionism associated with the statist variant was to be replaced by pro-market policy measures, such as deregulation, privatization, and liberalization. Unlike the statist variant, this strategy was more engaged at the international level. Free trade with the United States was for liberal continentalists the market solution to Canada's adjustment problem. By promoting cross-border integration and specialization of production, free trade would bring about productivity gains, growth of value-added activities, and expansion of the domestic technological base.

The switch to the international arena for problem solving occurred as Canada's relative position in the North American continent improved. Many began believing that Canada could negotiate an agreement with the United States that established institutional rules that would help ensure a more balanced division of positive gains for Canada, enhance Canada's political influence within the continental free trade framework, and help Canada manage the costs of its vulnerability dependence. Supporters of the liberal variant rebuffed the critics who argued that a bilateral free trade would inevitably undermine Canada's ability to assert its political autonomy and maintain economic sovereignty. Nossal (1985, 77) summarizes the logic underpinning the position of statist opponents:

> A wealthier Canada would be better able to assert its independence and sovereignty. With a more rationalized, competitive and productive manufacturing sector, and an ever-increasing real income, the Canadian people and their government could . . . better 'afford' the measures required for the maintenance of the nation's independence and sovereignty.

Canada's political-economic institutional system matched well with the institutional requirements of this adjustment strategy. Ikenberry (1988) and Katzenstein (1978) observe that states like Canada, whose political-administrative structure is fragmented and decentralized, whose federal structure lacks channels to foster intergovernmental coordination, and whose relations with business are detached or 'unembedded,' encounter huge obstacles if they seek to expand their interventionist capacity. But, the same states will be at an advantage if they seek to shrink the boundaries of state authority in the economy because they do not face network linkages that keep them entangled with societal groups. Moreover, the Canadian state, and others like it, can harness to greater effectiveness their commitment to market liberalism to sharpen market mechanisms, and they can apply the market principle to their federal system— that is, induce a coordinated process of competitive pro-market reforms among sub-national governments and between the two levels of government.

The ideas and tasks embodied in the liberal variant proved to be more consistent with the established mode of interaction between business and government. Until the 1970s, the policy routine of "embedded" liberalism helped promote free market enterprise, while legitimizing a limited economic role for the state. When the state expanded its economic involvement in the 1970s, the business community grew increasingly annoyed, viewing the government as "overly intrusive in attempting to regulate many areas of day-to-day business activity in ways that appeared to serve no discernable public interest" (Hale 2006, 151). Economic liberalism proposed a set of policy actions more in tune with the preferences of the business community.

The liberal variant of the Third National Policy was effective in harnessing the forces of competitive federalism to accomplish the objectives of this strategy. As an important institutional dynamic in the Canadian political order, competitive federalism was one of the principal institutional forces that had undermined the statist agenda because it had led both levels of government to inject greater state authority in the economy without an appropriate intergovernmental framework to coordinate their actions. Instead, the two levels of government pursued conflicting goals and intensified intergovernmental wrangling in a domain to which the constitution grants neither level exclusive jurisdiction.

Under liberal continentalism, competitive federalism would reinforce the process of state retreat from the economy that a free trade arrangement sets in motion. As Simeon (1986, 458) points out, "Any such treaty may well have the long-term effect not of strengthening either Ottawa or the provinces, but rather of constraining and limiting both, since the quid pro quo for the security of access we seek in the United States will undoubtedly be strong limitations on Canadian industrial and regional development policies, both federally and provincially." By exposing federalism to the demand of the market, competitive federalism would have a "market-preserving" effect to the extent that governments at both levels would compete to win the support of private sector actors by pursuing market policies that help maximize efficiency and profits. A government's policy success would compel other governments to emulate that policy and thus reinforce market mechanisms throughout the economy.

Conclusion

In the following chapters I will analyze the institutional situation associated with each strategy. The institutional situation corresponding to economic nationalism is characterized by a set of institutional dynamics that is resistant to institutional adaptation while the one associated with liberal continentalism features a set of dynamics that is receptive to adaptation. I maintain that the first institutional situation—institutional resistance—is denoted by three characteristics. The first characteristic of institutional resistance is the circuitous pattern in which the implementation of an economic adjustment strategy is pursued. Rather than a self-reinforcing pattern of implementation, institutional resistance causes government officials to resort to a series of discontinuous policy actions. Second, institutional resistance arises from the opposition of institutional actors as they seek to preserve their advantages derived from the existing institutional order. Lastly, and linked to the first two characteristics, institutional resistance is illustrated by the erosion of initial policy objectives and the tendency to settle on the lowest-common denominator solution.

The opposite results are found when liberal continentalism is implemented. First, the problem of adjustment is dealt with in accordance to a set of policies that becomes reproduced via self-reinforcing dynamics. Second, institutional receptiveness is represented by the actors independently adapting their behavior to the new path of policy development. Finally, institutional receptive-ness is illustrated when state actors are not forced to water down initial policy objectives; instead, they are able to build on initial policy moves, thus pushing policy development down a particular path.

Chapter Four

The Limits of Trade Diversification

Diversification was the economic nationalist strategy's way of seeking international redress of Canada's adjustment problem. The primary aim of diversification is to reduce a country's vulnerability to the actions of another country with which it is deeply linked economically and/or politically. When President Nixon announced in 1971 that the United States would seek to advance its national interests more overtly, Canada feared that relations with its neighbor would change for the worse. In response, Canada began developing counterweights in order to balance out the country's heavy reliance on the United States. Europe and Japan, in particular, became the focus of Canada's diversification policy.

This chapter focuses on the efforts of the Canadian government to pursue diversification in the 1970s and early 1980s. A central question guides the analysis: How and why did Ottawa's diversification policy change during this period? Ottawa's efforts to diversify via the contractual links with Europe and Japan were largely ineffective because the government could not persuade Canadian industries to alter their export habits in a way consistent with the goal of creating strategic counterweights. The reluctance of Canadian exporters was attributed to the enormous initial costs—transaction costs and information asymmetries—associated with doing business in relatively unknown, distant markets. Given the lack of export promotion assistance by Ottawa and of strategic coordination between business and government—two powerful tools of persuasion—Canadian exporters had little incentive to diversify trade.

The failure of the contractual links led Ottawa to renew its efforts to seek a favorable multilateral agreement in the Tokyo Round of the General Agreement on Tariffs and Trade that began in 1973. Recognizing the rising public and governmental support for the multilateral track, Ottawa established a consultative and coordinative framework that provided voice opportunities to the provinces and societal groups in order to better represent Canada's interests during the Tokyo Round. The shift toward renewed multilateralism meant that Canada's efforts to minimize its economic dependence on the United States would not be achieved by creating strategic counterweights, but by subscribing to the principle of 'safety in numbers', a formula whereby "power matters less, rules matter more, and supportive coalitions can be engineered" (Stairs 1995, 54). Canada's confidence in the success of this strategy also stemmed from the fact that Ottawa had experience in the art of multilateral trade negotiations.

The Thrust toward Diversification

Canada had never been too vulnerable to American actions until President Richard Nixon announced in 1971 a series of economic measures that threatened to disrupt the bilateral economic relationship. The concern over the tendency of not doing anything about Canada's economic dependence on the United States had been addressed in *Foreign Policy for Canadians*, a white paper published in 1970 that reported the results of a special review of Canada's foreign policy that was launched in May 1968 by Prime Minister Trudeau. The white paper (1970, 38-9) summed up Canada's international situation as follows:

> Canada's particular situation, [the fact that it is in the shadow of American economic and political influence] requires a certain degree of self-reliance and self-expression if this country is to thrive as an independent state. The key to Canada's continuing freedom to develop according to its own perceptions will be the judicious use of Canadian sovereignty whenever Canada's aims and interests are placed in jeopardy.

No major foreign policy changes, however, were made in response to the white paper. The import tax surcharge that the United States imposed in August 1971, from which Ottawa sought unsuccessfully to be exempted, provoked strong reactions in Ottawa.[1] The 'Nixon shocks' demonstrated how risky it was for Canada to depend on the goodwill of the United States, especially when the American market took in about 65% of Canada's exports at the time. As Thomas Axworthy (2000, 79), who at the time was the Principal Secretary to Prime Minister Trudeau, pointed out, Canada's anxiety was the consequence of the fact that "most of the trade eggs were in one large basket."

---

1. During these bilateral exchanges, the Canadian government, not knowing what the impact of the United States' economic measures would be on the Canadian economy, tactically used inflated figures to try to soften the Americans into granting Canada an exemption. In the end, Nixon's measures did not inflict severe costs on the Canadian economy and eventually accommodations were made between the two governments.

Officials in Canada were firmly of the view that their country had least contributed to the economic problems the United States was seeking to resolve. A day after Nixon announced his economic program, then Secretary of State for External Affairs Mitchell Sharp remarked that if the measures "are supposed to induce countries to change their unfair exchange rates or to remove discriminatory restrictions against American imports . . . we're not guilty on either count" (quoted in Dobell 1985, 15). In responding to the American suggestion that Canada should limit itself to exporting natural resources to the United States, Sharp forcefully stated:

> Canada . . . has built a balanced and successful industrial and trading economy. I can assure you that Canada is determined to continue on the course it has set for itself. Suggestions from responsible authorities in the United States that Canada should reduce its secondary manufacturing industry and concentrate on the exploitation and processing of natural resources are as insensitive as they are uninformed (quoted in Dobell 1985, 19).

Recognizing that Canada would not receive an exemption, the Canadian cabinet toward the end of August 1971 requested that Sharp assess the state of the bilateral relationship and explore the feasibility of alternative policy options. Toward the end of September, Sharp submitted a memorandum to the cabinet, in which three options were identified as possible strategies to guide Canada's policy toward the United States. The three options were: first, "to maintain more or less our present relationship with the United States;" second, "move deliberately toward closer integration with the United States;" and third, "pursue a comprehensive, long-term strategy to develop and strengthen the Canadian economy and other aspects of our national life and in the process to reduce the present Canadian vulnerability" (Sharp 1972, 13). After weighing each option, the cabinet chose the third option.[2]

The first and second options were rejected on the grounds that each risked pushing Canada closer into the American economic orbit and did not offer solutions to the problem of vulnerability. Moreover, while the options paper made clear that the first two options made economic sense, they were rejected on the grounds that Canada lacked the necessary power leverage to change the pattern of bilateral economic ties in its favor.

The "Third Option" proposed that a policy of diversification was Canada's best foreign policy option to deal with its adjustment problems. Two factors shaped Ottawa's views on diversification. First, diversification was motivated

---

2. The options paper authored by Sharp never assumed the status of a government white paper. Instead, released in the 1972 autumn issue of the journal *International Perspective*, published by the Department of External Affairs, the paper served more as a think-piece among government officials and the public, and the paper's endorsement of the third option reflected the intellectual discourse within Trudeau's Liberal government.

more by political calculations than by economic ones; its aim was to safeguard Canada's political independence and strengthen its economic autonomy. Ottawa under-stood that the policy would not bring about economic disentanglement of the two countries; instead it was hoped that the policy would result in balancing out the asymmetry of bilateral interdependence via the cultivation of counter-weights. As Sharp pointed out in 1972 before the Chamber of Commerce in Buffalo, New York:

> For Canada, there is not, and will not be, any substitute for the market this country represents. Canadian prosperity depends on access to the American market. But I think that, if there is one thing Canadians and Americans agree about, it is that Canada should remain free, sovereign, and independent. Diversification of relations does not imply disengagement from our community of interest with the United States.

The second factor guiding the thinking of policymakers on diversification was the recognition that the policy needed to be pursued systematically over a long period of time. Diversification, as Stairs (1999, 233) observes, was an "act of will, in defiance of fate," or as Bothwell (1977, 35) remarks, a determined "attempt to secure the triumph of politics of geography." Previous attempts to diversify Canada's economic ties had produced meager results. For example, little came of Prime Minister John Diefenbaker's 1958 national development policy to divert 15% of Canada's trade from the United States to Great Britain, and increasing trade with the Commonwealth (Hart 2002). If past trends offered any advice to policymakers involved in the implementation of the diversification policy it was that creating counterweights for the purpose of generating countervailing influences to the strong American economic embrace would "take time to materialize," require "a conscious and deliberate effort" in the application of statecraft in order to "put and maintain the Canadian economy on such a course" (Sharp 1972, 17).

Linking Up With Europe and Japan

Atlanticism has always influenced Canadian foreign policy. For some, the Atlantic idea refers to the notion that Canada and Europe share historical ties and common values; for others, it alludes to the triangular relationship linking Canada, Europe, and the United States through which common interests are pursued; and still for others, Atlanticism refers to Canada's quest to project its influence and secure its relative position within those institutions that bind states on both sides of the Atlantic (Nossal 1992). In short, although Canada has been drawn ever more deeply into the American orbit over the course of the twentieth century, Europe has always figured prominently in Canadian foreign policy.

It came as no surprise therefore when the White Paper homed in on Europe as a potential counterweight. As a basis for deepening their economic ties,

Canada attempted to draw a parallel between Europe and Canada regarding their vulnerability to American policy actions. In the European portion of the White Paper (1970, 14), the government observed that "[t]he maintenance of an adequate measure of economic and political independence in the face of American power and influence is a problem Canada shares with the European nations, and in dealing with this problem there is at once an identity of interest and an opportunity for fruitful cooperation." But there was another factor that pushed Canada to strengthen its economic relations with Europe. Britain's entry into the European Community in 1973 was of concern because it doubtless would affect Canadian relations with its European partner. After studying the impact of British entry into the European Community, the Senate Standing Committee on Foreign Affairs (1973, 2) reported that such a move "would throw a greater burden of adjustment on Canada than on any other country outside the enlarged EC." It was estimated that over $600 million of Canadian exports to Britain would be adversely affected overtime due to the elimination of the preferential tariff arrangement Canada had with Britain.

During the five years it took to arrive at the Framework Agreement for Commercial and Economic Cooperation with the European Community in July 1976, Canada insisted on attaining three key objectives. First, Canada sought to create an institutional framework that would facilitate closer economic ties with the European Community. However, there was much vagueness about the substance of the agreement Canada was seeking, a factor that contributed to Europe's lack of interest and polite indifference to Canada's diplomatic initiative. Without a 'big idea' to catch the attention of Europe, Canada made several unsuccessful diplomatic overtures before arriving at a design that gained the support of the European Community (Mahant 1981, 272).

The second objective was to obtain a collaborative arrangement that enhanced Canadian opportunities to increase exports of secondary manufactured goods and increase industrial cooperation with Europeans. Central to this goal was the desire to dispel the view among European governments that Canada should serve as Europe's supplier of natural resources. Ottawa sought an arrangement that did more than produce mutually positive gains; it sought to secure provisions that would directly assist domestic efforts to strengthen Canadian manufacturing capacity. Lastly, Canadian negotiators insisted that Europeans recognize Canada as having interests distinct from those of the United States. During a conference with representatives of the EC in Ottawa in November 1972, Sharp noted:

> Outside this country, I have sometimes found an assumption that Canada should fall naturally and inevitably into the U.S. orbit. This is perhaps understandable, but it is unacceptable to Canadians. It is contrary to the Canadian Government's basic policy of a relationship distinct from but in harmony with the United States. Canada's relationship with Europe is not the same as the United States' relationship with Europe. Perhaps, in relative terms, our relation-

ship is more important to us than the United States' relationship with Europe is to the Americans.

To Ottawa this was important because, indeed, there were real differences of interests between the United States and Canada. But more so, Ottawa wished to avoid finding itself in an embarrassing situation where the European Community felt the need to consult with the United States before it agreed to enter into a cooperative arrangement with Canada. For Canadian negotiators, it was important to make clear to Europeans that how Ottawa conducted its foreign economic policy was the concern of none other than Canadians.

Negotiations started slowly. In December 1971, it was agreed by both sides that bilateral consultations would take place twice a year. As petitioner, Canada faced two initial obstacles in its efforts to gain the interest of Europeans in strengthening bilateral ties. First, although trade between the two partners was of some significance, the Europeans needed the United States more than they needed Canada. Consequently, some European officials began wondering what effect an EC-Canada agreement would have on Europe's relationship with the United States. This concern came to a head when reports surfaced suggesting that Washington was pressuring some European governments to block Canadian proposals.

Second, Trudeau's 1969 decision to cut the number of Canadian troops stationed in Europe under the umbrella of the North Atlantic Treaty Organization signaled to European capitals that Canadian relations with Europe had become less important to Canada's national interest. Thus, when Ottawa expressed an interest in a commercial arrangement with the EC, European officials were initially confused. Subsequently, the position of European officials was that if Europe was indeed important to Canada, then Ottawa had to demonstrate it by being involved not just economically, but militarily and politically as well.

An opening occurred in November 1973 when the European Community requested that Canada make its intentions known on the nature of closer economic ties. In April 1974, the Canadian government delivered a proposal calling for the initiation of negotiations "with the appropriate Community institutions with a view to concluding a trade agreement [that would] effectively underpin the contractual relationship with the Community which is currently based on common adherence to the General Agreement on Tariffs and Trade" (quoted in Granatstein and Bothwell 1990, 164). The proposal received a cool reception in Europe. Two factors contributed to this outcome. First, the Europeans regarded Ottawa's proposal as being too vague for their taste and feared that the proposal would undermine the General Agreement on Tariffs and Trade framework even though Canada underscored that the contractual relationship would comply with the rules and principles of the multilateral trade regime. Second, in light of the energy crisis of 1973, the Europeans were more interested in making the goal of stabilizing the supply of energy the basis for pursuing closer bilateral economic relations.

Ottawa, however, was not prepared to give Europeans access to Canadian natural resources. Conveying this point after a visit to Europe in 1974, Trudeau remarked that "we are telling the Europeans bilaterally and as a community: you may think you are going to be able to take all our raw materials out, but you aren't" (quoted in Ryan 1974, B2). Nevertheless, the Prime Minister was willing to employ resources as a bargaining chip to gain the acquiescence of reluctant Europeans. "We are defining our energy and natural resources policies and if you want to get in there, you'd better embark on this process of negotiations," Trudeau pointed out.

Moreover, to strengthen their case, Canadian negotiators were not shy about conveying what might happen if negotiations failed. As Kinsman (1973, 25) put it, "The alternative to the success of Canada's third option could be a North American economic bloc dominated by the U.S. economy that would . . . reduce the possibilities of EEC access to Canadian industrial markets and energy resources in conditions of increasing scarcity." If anything good came of these diplomatic exchanges it was that the Europeans were beginning to take Canada more seriously. Aside from the effective pressure tactics, Ottawa's renewed interest in NATO by 1973 certainly figured in Europe's change of view toward the diversification policy of Canada.

During a visit to London in March 1975, Prime Minister Trudeau unveiled a plan with which the Europeans were more content, or at least, which they were willing seriously to consider. Up to now, although there was a recognition of the opportunities for achieving common interests, differences over how to translate them into an agreement prevented the two sides from making headway. Trudeau (1975, 4-5) laid out the logic underpinning the recent proposal:

> [It is] the desire of Canada to enter into a contractual relationship with the Community—one that would ensure that both the Community and Canada would keep the other informed, would engage regularly and effectively in consultation, would not consciously act to injure the other, would seek to co-operate in trading and any other activities in which the Community might engage. Because we do not know . . . how far or how fast [Europe's] experiment in integration will take it, or what form it will assume on arrival, no overall agreement can be laid in place at this time. But what can be done is to create a mechanism that will provide the means (i.e., the "link") and the obligation (i.e., "contractual") to consult and confer, and to do so with materials sufficiently pliable and elastic to permit the mechanism to adapt in future years.

On June 25, 1975, in a unanimous decision, the Council of Ministers gave a mandate to community officials to negotiate a framework agreement with Canada. However, Britain and France were less enthusiastic about proceeding with the negotiations because they were concerned about preserving their right to maintain and deepen their individual economic ties with Canada. By October, however, Britain and France had come around thanks to a promise that a Community agreement with Canada would not prevent individual members to deepen bilateral relations of their own with Canada.[3] The negotiations were brought to a successful conclusion in early June and culminated on July 6, 1976 in the signing of the Framework Agreement for Commercial and Economic Cooperation.

The Agreement was the product of Canadian efforts to institutionalize a consultative mechanism to facilitate economic cooperation with the European Community. The Joint Co-operation Committee (JCC), established under Article IV of the Framework, was given the mandate "to promote and keep under review the various commercial and economic co-operation activities envisaged between the Communities and Canada." Allan MacEachen, then Secretary of State for External Affairs, remarked during the signing of the Framework that the JCC would be in charge of infusing the agreement with life. In particular, the JCC was to play an important role in helping the business communities of Europe and Canada cultivate broader interfirm links, such as joint ventures; enhance their involvement in each other's industrial development endeavors; and increase the exchange of technology and scientific knowledge.

Moreover, the Framework represented a collaborative arrangement that ensured a balanced division of mutual gains between the European Community and Canada. Canada's desire to export a full range of products—semi-processed or fully-processed primary or secondary products—had been heeded by Europeans. During a speech before the Canadian Institute of International Affairs within a few months of the signing, Marcel Cadieux, who at the time served as Canadian ambassador to the EC, remarked that a mature relationship underpinned the bilateral agreement and that "classical nineteenth-century deals involving the mere export of unprocessed raw materials are behind us."

Canada's contractual link with Japan resembled the one struck with the European Community. Just as the Canada-EC agreement was owed to Canadian diplomacy and Trudeau's personal involvement, so was the agreement between Canada and Japan. It took sustained efforts to persuade Japan to purchase more Canadian manufactured goods and to undertake more capital-intensive investments in Canada. During the Pacific Basic Economic Cooperation Council meeting in Vancouver in May 1971, the Minister of Industry, Trade, and Commerce, Jean-Luc Pépin, urged Japan to purchase more Canadian fully manufactured and high technology goods as well as encouraged Japanese businesses to increase their participation in Canada's processing and manufac-

---

3. In response to their concern, section four of Article III of the Framework Agreement for Commercial and Economic Cooperation states that "the present Agreement and any action taken thereupon shall in no way affect the powers of the Members States of the Communities to undertake bilateral activities with Canada in the field of economic co-operation and to conclude, where appropriate, new economic co-operation agreements with Canada."

turing activities using local supplies and equipment and Canadian management. During a visit to Tokyo in September 1973, the new Minister of DITC, Alastair Gillespie, conveyed the same message, stressing the need for greater access to Japanese markets and Japanese investment in Canada (Langdon 1980).

A few months after the signing of the Canada-EC commercial framework, Ottawa came within reach of a similar agreement with Tokyo. On October 21, 1976, the two sides signed the Framework for Economic Cooperation. A Joint Economic Committee, headed by the foreign ministers in both countries, was created and given the mandate to expand bilateral trade, increase investment, and encourage technological and scientific exchanges (Granatstein and Bothwell 1990, 174).

Act of Will, But Not in Defiance of Fate

Political will coupled with effective statecraft are what enabled Canada eventually to establish framework agreements with Europe and Japan. However, the weakness of domestic institutions and the force of geography rendered the diversification plan ineffective. From the beginning, Canadian officials recognized that the success of diversification would depend on Ottawa's ability to get the business community aboard. However, efforts to mobilize business collaboration behind the diversification plan were undermined by several factors: the arm's-length pattern of business-government relations, the decentralized structure of business, the limited resources that Ottawa could bring to bear on export promotion, and bureaucratic struggles among executive departments. Reflecting on this chapter of Canadian economic adjustment measures, Axworthy (2000, 80) observes that changing the "individual decisions of thousands of exporters [was] beyond the [government's] power. [T]he ability of government to redraw trading patterns was too limited to turn the diversification goals of the Third Option into a reality."

The lack of consultative and coordinative efforts between Ottawa and the business community, as Ottawa negotiated each agreement, reflected the institutional realities between the two actors—that is, a policy legacy of limited government interface with the business community. The few consultative measures that took place occurred in an ad hoc fashion. The view in Ottawa, however, was that if negotiators produced a favorable bilateral agreement, then Canadian entrepreneurs would automatically use its provisions to expand their business to Europe and Japan. A few months after signing the Framework Agreement with the European Community, Ambassador Cadieux (1976, 5) remarked that having laid down an economic framework "it will be up to the entrepreneurs, the investors, the financiers, the bankers to see what can be done to develop our relations with Europe."

However, Ottawa did take certain measures designed to create positive inducements. For example, section three of Article III of the Framework Agreement was designed to induce Canadian business to increase trade with

Europe by facilitating the exchange of information regarding commercial opportunities and encouraging such economic activities as joint ventures, direct investment, exchange of expertise and technological information with their European counterparts. But it became clear that this provision had a minimum effect in helping Canadian exporters to overcome the problem of information asymmetries and transaction costs. This led the Department of Industry, Trade, and Commerce to arrange a series of seminars under the title "Perspective on Western Europe" that were offered in conjunction with Enterprise Canada 77, a consultative initiative aimed at achieving a closer relationship between the private sector and the government. However, the initiative generated little change in the export habits of Canadian firms.

The decentralized structure of business and the geographical appeal of the American market combined to dissuade Canadian entrepreneurs from taking advantage of the opportunities offered in the Framework Agreements. Even before the Framework Agreement was signed with Japan, it was clear that Canadian business lacked the organizational capacity to mount a strategy to sell Canadian products to Japan. Langdon (1980, 79), for example, observes that during a trade mission led by Pépin in January 1972, in which the objective was to sell nuclear reactors and aircrafts, "the businessmen . . . were not organized and prepared for the sort of high-powered approach being urged on them by the Cabinet minister." As a result, both the Canadian government and the Japanese businessmen undertook the initiative to complete the transaction.

Another factor influencing the behavior of Canadian firms was the differential costs associated with exporting to Europe and Japan versus the United States. The interaction between Canadian and American firms, dating back to the early 1900s, had the effect of creating similarities in business practices, from marketing to production methods to technology use to management systems. Decades of cross-border business contacts and bilateral economic ties had contributed to the development of two national systems of economic institutions that were highly compatible with one another, which in turn had contributed to lowering transaction costs and solving the problem of information asymmetries. As the Economic Council of Canada (1975, 97) pointed out in a 1975 report: "Canadians are better able to develop systems of business operations in the United States than in Europe, because they are more attuned to U.S. institutions, practices, and even habits of thought."

The lack of export promotion capacity of the federal state also contributed to the reluctance of the Canadian business community to take advantage of the two Framework Agreements. Granatstein and Bothwell (1990, 176) remark that "corporations had waited for Ottawa to offer incentives to encourage their efforts in Europe and Japan, but there were no tax breaks, no export incentives, and no subsidies to encourage research and development aimed at world markets." This view was also echoed in the *Final Report of the Export Promotion Review Committee*, referred to as the Hatch Report, named after the committee chairman, Robert Hatch. The committee, whose seventeen members all came from the private sector, was established in December 1978 by the Minister of Industry, Trade, and Commerce to assess the effectiveness of the government's past and current export promotion programs.

In its report, the committee remarked that government services such as export marketing and promotion, export financing and insurance, the trade-foreign aid interface, and export taxation policy were inadequate. As the Hatch Report (1978, 13) pointed out: "Relative to other leading nations, not only is business-government coordination deficient, but also coordination between the various federal agencies that represent Canada abroad and between federal and provincial governments." With respect to the Export Development Corporation, which was established in 1969 to provide finance and insurance to assist Canadian export trade, committee members had this to say:

> At the heart of the [export promotion] problem is the fact that the [Export Development Corporation] does not, cannot and will not subsidize exports. It simply is not set up for it. Of greater importance are criticisms about the lack of international competitiveness in medium-term financing, the absence of short-term financing for small exporters, the lack of involvement of the commercial bank. If Canada is to move aggressively to increase [its] exports, it is essential that we realize how the competitive game is played by other nations (Hatch Report 1978, 28).

The other factor that undermined the policy of diversification was the clash of interests between the Department of External Affairs on the one hand and the Departments of Finance and of Industry, Trade, and Commerce (DITC) on the other. The 1970 White Paper had emboldened the Department of External Affairs to expand its field of interests into the economic domain of Canada's foreign policy, a policy field that had traditionally been led by Finance, the Bank of Canada, and the Department of Trade and Commerce (until it was merged with the Department of Industry in 1969 to create the Department of Industry, Trade and Commerce).

Officials in Finance and the DITC forcefully resisted the expansionist ambitions of External Affairs (Keenes 1992). Moreover, Finance officials, in particular, rejected the counterweight logic because its policy design intruded in business affairs by pushing on them a politically motivated design, shifted attention away from the more important concern of enhancing Canadian international competitiveness, and deviated from the traditional multilateralist track that had been credited for bringing about postwar economic growth and industrial development. Without the full support of Finance and DITC, External Affairs was constrained in its ability to boost Ottawa's export promotion capacity behind diversification since the two departments had at their disposal important promotional instruments.

Toward Renewed Multilateralism:
The Tokyo Round

Multilateralism has had a longer following in Canada than has diversification. In the context of the Canadian relationship with the United States, multilateralism and diversification have promoted Canadian goals by deflecting American influence and managing Canadian vulnerability. As former Prime Minister Lester Pearson once pointed out, Canada's involvement in multilateral settings has "helped us escape the dangers of a too exclusively continental relationship with our neighbor without forfeiting the political and economic advantages of that inevitable and vitally important association" (quoted in Keating 1993, 20). The main difference between the two policy thrusts rests on the mechanism each employs to attain national goals. Multilateralism has been based on the formula of safety in numbers that is exemplified by the structure of international organizations—such as the General Agreement on Tariffs and Trade (now the World Trade Organization), the North Atlantic Treaty Organization, the Organization for Economic Cooperation and Development, and the United Nations and its affiliated international organizations—whereas diversification has been based on the selective cultivation of counterweights (Stairs 1994).

As a rules-based multilateral trade regime, the GATT/WTO has well served Canadian interests. For starters, the regime's push for greater trade liberalization has created export opportunities for Canadian producers. In the postwar period, the continuation of multilateral reductions of tariffs became imperative for Canada because an increasing share of its gross domestic product was generated by exports. Another advantage that Canada has enjoyed from the multilateral trade regime relates to the regime's set of international rules and decision-making procedures. The international rules embodied within the regime have constrained and shaped the commercial policies of member states, as illustrated by restrictions on the use of various trade protective tools. Moreover, the regime's decision-making procedures have allowed Ottawa to voice its interests and concerns. Finally, Canada has benefited from the GATT/WTO's dispute settlement mechanism. Despite its flaws, the multilateral dispute settlement process has given member states a mechanism through which they can seek redress of economic disadvantages caused by the protectionist measures and unfair trade practices of other member states.

Initially, the "Third Option" did not consider the Tokyo Round of multilateral trade negotiations, launched in 1973, as a vehicle for advancing the goals of the Third National Policy. However, as it dawned on government officials that the contractual links would bring about limited trade diversification, Ottawa renewed its focus on the multilateralist track, zeroing in on the Tokyo Round. Despite the inopportune circumstances surrounding the launching of the Tokyo Round of multilateral trade negotiations—the 1973 oil crisis, the increasingly aggressive economic policy of the United States, the international economic slowdown, and European regionalism—by its conclusion in 1979, trading nations had achieved some major breakthroughs and Canada had agreed to make some substantial cuts in its tariff levels.

Until the Tokyo Round, Canada had been one of the few industrialized nations—Australia and New Zealand were the other two—to have reduced its tariffs on dutiable imports at a much slower rate than other industrialized nations. Tariff protection was a key instrument of economic policy in Canada dating back to the National Policy in 1879. Originally championed by the Conservative Party and the business community in central Canada and opposed by the natural resource-based economies of the western and eastern provinces as well as the Liberal Party, by the 1940s there was overwhelming societal and political support for tariff protection.

In the Kennedy Round, Canada opposed using the proposed linear or across-the-board tariff cuts formula, which called for the reduction of tariffs on dutiable industrial products by 40%. Instead, it opted to use an item-by-item approach to tariff reduction with individual trading partners. This decision was based on the observation that under the linear formula Canada would obtain an unequal distribution of gains. At the time, Canada was a net importer of manufactured goods, maintaining relatively high domestic tariffs in this category of goods, as well as a net exporter of primary goods, which faced low international tariff levels. The proposed reduction in tariff levels on dutiable manufactured goods would have increased import competition in a group of goods where Canada faced a comparative disadvantage without offering offsetting international tariff cuts in a category of goods where Canada maintained a comparative advantage (Sharp 1995). In the end, the Kennedy Round delivered some notable gains to Canada. Commenting on the outcome, Evans (1971, 54) observes: "Canadian negotiators managed a remarkable 'tour de force'. . . . Over three billion in Canadian exports stand to benefit . . . while Canada has agreed to reduce on some two and a half billion of imports [which means that] nearly half [of] Canada's present dutiable imports remained untouched."

What factors explain Canada's decision to move to a low tariff policy starting in the Tokyo Round? The unwillingness of major economic powers to extend to Canada during this Round of negotiations the same concessions they had agreed to in previous rounds was certainly an important factor contributing to this policy change. But three other factors stand out as well. First, although the policy of strategic diversification and multilateralism were pursued simultaneously in the 1970s, Ottawa's inability to overcome domestic obstacles associated with the former ruled out hopes by 1977 that the policy could produce some significant change in the realities of the bilateral relationship. Still determined to use the international trade system to diversify economic ties, Ottawa renewed its interest in multilateralism, proposing a major change in its commercial policy.

If the goal of the "Third Option" was to prevent further Canadian export dependence on the United States, then the policy of diversification was clearly the optimal policy, for it sought deliberately to establish supplementary markets in order to diversify Canada's exports away from the United States. Canada's renewed interest in multilateralism, however, was a second-best choice because in addition to promoting international liberalization (and increasing market access abroad), the results of past multilateral trade negotiations had drawn Canada closer to the United States. Stone (1984, 179) notes that Ottawa defended its position, arguing that diversification would fail without the multilateralism of GATT: "In discussing the Third Option concept, Mr. Sharp dismissed any suggestion that the liberal world-trade system was responsible for the imbalance in Canada's trade relationship with the United States; indeed, he expressed the view that a less liberal world-trade system would have led to even stronger links between the U.S. and the Canadian markets."

However, contrary to Sharp's assertion, the trade statistics of the post-Dillon Round of the GATT (1960-1) showed that the share of Canadian exports destined for the United States had increased from 60% in 1961 to 70% in 1969. The pull of geography was the same after the Kennedy Round as Canada saw its proportion of American-bound exports increase from 70% in 1970 to 80% in the early 1980s. Thus, although Ottawa was firmly of the view that diversification could be achieved through multilateralism, the realities clearly showed that after the conclusion of trade reduction negotiations, Canada grew more trade dependent on the United States.

The second factor, related to the first, is that Ottawa began seriously to question the merits of tariff protection, especially in the face of pressures to attain economic adjustment. Canada had experimented with freer trade in the 1960s and the results were positive. The 1965 Canada-United States Automobile Products Trade Agreement put in place institutional terms allowing for duty-free trade in cars, trucks, buses, parts, and automobile-related accessories. Moreover, the Auto Pact devised provisions that served as incentives for automakers to increase their manufacturing activities in Canada. With the Auto Pact in place, Canada witnessed not only an increase in the production of automobiles—as a result of specialization and productivity improvements—but also a change in the composition of Canadian exports, for automobiles and related products began to make up a larger share of export commodities. Finlayson and Bertasi (1992, 25) observe that as a share of merchandise exports bound for the United States, manufactured goods increased from 15% in 1965 to 30% in 1970, an increase owed largely to the surge in exports of automobile and parts. The positive results of this sector-specific free trade experiment set off a riveting debate in the 1970s, which attacked the conventional wisdom that Canada needed tariff protection in order to build and preserve its manufacturing capability.

Finally, the policy change was coincident with the change in Canada's trade profile. Starting in the mid-1960s, Canada saw an increase in exports of manufactured commodities, partly due to the Auto Pact and to exports in other categories of end-products. In their review of Canada's trade position in the second half of the 1970s, officials at the Department of External Affairs (1983, 7) observed that "Canada had developed a diversified industrialized base capable of exporting a wide variety of products to markets all over the world." Accordingly, Canadian negotiators could no longer make a credible argument that Canada warranted preferential treatment because it had a weak manufacturing base and its economy was primarily resource-based. Mounting domestic and international pressures forced Ottawa to abandon the practice of free riding on others' tariff cuts, and to demonstrate its commitment by equally contributing to multilateral tariff reduction.

As the policy of diversification illustrated, political will does not assure effective policy implementation. Therefore, it is important to consider what conditions contributed to the success of renewed multilateralism. Whereas Ottawa was constrained in its ability to mobilize societal and bureaucratic consensus behind diversification, such was not the case with renewed multilateralism. Moreover, Ottawa actively promoted consensus-building and coordinative activities with the provinces.

Canada entered the Tokyo Round with several objectives. Ottawa sought to reduce the levels of tariffs and non-tariff barriers that Canadian exporters faced in foreign markets, restrict the arbitrary use of trade protective instruments by foreign governments against Canadian exports, and establish specific terms that would help spur development of more sophisticated forms of Canadian processing of primary and secondary manufactured goods. Moreover, Canada wanted to enhance the international competitiveness of Canadian industries by lowering Canadian tariff levels, while limiting tariff concessions in employment sensitive industries and allowing for orderly adjustment by private actors (Roy 1984).

A major preoccupation of Canadian negotiators once the Tokyo Round was inaugurated in 1973 concerned the formula that members would agree to use in order to coordinate this Round of tariff reductions. From the outset, Canada favored using a sector approach, one that held the greatest hopes of securing Canadian objectives. The sector approach, as Winham (1986, 237-8) points out, was "originally proposed by Canada . . . for negotiating all tariff and nontariff barriers in selected commodity areas, and particularly including goods at different levels of processing." Although this formula was preferred by resource producing nations, it failed to gain the support of major economic powers because they believed that the sector approach did not generate the kind of reciprocity they were hoping to receive.

It was not until late 1977 that the major economic powers—the United States, Japan, and the European Community—settled on a tariff-cutting formula referred to as the Swiss formula. Unlike the sector approach, the Swiss formula called for a 30%-40% cut in tariff levels across industries and proposed deeper cuts for countries with higher tariffs than those with lower ones. In January 1978, Ottawa announced that it would accept this formula. With the conclusion of the multilateral trade negotiations, the depth of cuts for all industrial products that Canada agreed to was 38%, more than the 31% and 29% agreed to by United States and the European Community respectively, but less than Japan's 49% (Stone 1984, 182).

Several factors helped Ottawa mobilize cooperation within and outside the political system. This second attempt at international economic adjustment benefited from the fact that multilateralism had been used repeatedly in the postwar period to promote industrial development. By the 1970s, multilateralism had become a policy routine, backed by government officials at the provincial and federal levels and by the business community. Moreover, although multilateralism at this juncture was, to a certain extent, under statist influence, the fact that it helped expand market mechanisms and preserve the autonomy of market players made this policy all the more appealing to actors opposed to the strategy of economic nationalism.

The government employed three institutional innovations that were central to the success of renewed multilateralism: rational management within the federal state, government-business consultation, and federal-provincial consultation. Led by a senior bureaucrat at the deputy minister level and assisted by a small secretariat, the Canadian Trade and Tariffs Committee (CTTC), created in 1973, functioned as a clearing-house where the private sector and the provinces could submit briefs, which were subsequently distributed to the Departments of Finance, of Industry, Trade, and Commerce, and of External Affairs for review and gathering of information. In an effort to foster administrative coherence and coordination within the federal state, an interdepartmental Trade Negotiations Coordinating Committee (TNCC) at the level of deputy minister was established in the months following the inauguration of the Tokyo Round and given the task of preparing "recommendations to Cabinet with respect to Canada's objectives in the Tokyo Round, and to convey Canada's instructions to the trade delegation in Geneva" (quoted in Winham 1978-9, 77).

By 1977 there was significant pressure to revamp these coordinative mechanisms. Two sets of pressures had arisen. There was growing recognition that the provinces needed to be more involved in the process of international trade policymaking. Although international trade negotiations and the administration of tariffs are two responsibilities that fall under the authority of the federal state, the constitution equips the provinces with two important sources of trade authority. First, the provinces have the means to employ non-tariff instruments—such as regulations, subsidies, restrictive standards, and government procurement—and consequently can impede interprovincial and international trade. Moreover, the federal government is limited in its ability to implement treaties that have provisions that fall within provincial jurisdiction (de Boer 2002). When Ottawa cannot assert effectively its international competence because of jurisdictional issues, it needs to consult the provinces to ratify a treaty. Accordingly, if Ottawa wanted to tackle the issue of non-tariff barriers in the Tokyo Round, it needed to bring the provinces into the negotiation process. Furthermore, if the provinces, particularly western provinces, wanted to bring about further trade liberalization in Canada, they needed greater access to decision makers in Ottawa.

The second source of pressure arose from the need to increase further coordination within the federal state—in particular, within the federal bureaucracy, between the bureaucracy and cabinet, and between the negotiating team in Geneva and Ottawa. The TNCC fell into disuse soon after it was created because its organizational design matched poorly with the task it was intended to fulfill. As Winham (1978-9, 87) points out, "the TNCC was too oriented to the external negotiation and unable to generate the domestic political and preparatory work needed to support the Canadian position at the Tokyo Round." Ottawa needed to centralize its negotiation authority while establishing consultative and informational linkages that mobilized bureaucratic consensus and coordinated their actions.

In response to these pressures, the position of Canadian Coordinator for Trade Negotiations (CCTN) was created in August 1977. Upon receiving the appointment to head the Office of the Coordinator, Jake H. Warren discussed the tasks with which he was entrusted:

I am to be the key link between Canadian negotiators in Geneva and the cabinet and departments in Ottawa. I am to liaise with the provincial governments and with interested organizations in the private sector during the whole course of the trade negotiations, and I am responsible for coordinating for cabinet the preparation of the Canadian positions for all of the various elements of the negotiations (quoted in Winham 1978-9, 79)

The Coordinator became the linchpin of Ottawa's efforts to cultivate greater coordination. Warren served as a secretariat for the ad hoc Cabinet Committee on the multilateral trade negotiations (MTN), where authority in formulating Canada's MTN position formally resided, and he assumed the chair of the Federal-Provincial Committee, which oversaw intergovernmental coordination activities. The Coordinator also co-chaired the Interdepartmental Committee on Trade and Industrial Policy, a newly created committee in charge of addressing the problem of economic adjustment derived from the post-Tokyo Round tariff reductions.

The Office of the Coordinator helped build consensus among various actors not accustomed to working together. It helped centralize channels of communication within the federal state and provide clear direction to the negotiating team in Geneva. Moreover, the Office of the Coordinator helped improve federal-provincial relations in a manner unseen before in multilateral negotiations. As Winham (1978-9, 87) observes: "When the MTN began, the federal government treated the provinces as constituents[; however, by] the time major decisions had been made, the provinces have been advanced to the role of joint policy-makers, and on some matters they appeared to be as influential as the major federal departments."

Conclusion: From Renewed Multilateralism toward Continentalism

The rise of continentalism in the 1980s as a form of international redress of the adjustment problem was partly the product of the success of renewed multilateralism. Renewed multilateralism, as it turned out, paved the way for the shift to continentalism by locking in two key economic forces. First, renewed multilateralism reinforced Canadian reliance on trade. In 1980, exports accounted for 25.6% of Canada's gross domestic product, up from 19.5% in 1970 (Department of External Affairs 1983, 20). As Canada became more integrated in the international economy, Ottawa was pressured to change the way in which it addressed the national power consequences of its economic activities. Because an increasing proportion of Canadian national wealth was derived from international trade, there were higher stakes facing Canada in seeking a more open and rules-based international trade regime as well as in obtaining better terms of access to the markets with which Canada traded the most.

Moreover, in the 1970s the share of Canada's total exports destined for the United States reached unprecedented levels, fluctuating between 63% in 1970 and 70% in 1978 (Department of External Affairs 1983, 204). If the post-Kennedy Round effect of intensifying Canada's export market concentration was repeated after the Tokyo Round, as many believed would happen, an even higher level of Canadian exports would be destined for the American market in the 1980s. With Canada's trade pattern becoming increasingly bilateral and with trading accounting for a larger share of GDP, it was a matter of time before the issue of developing a bilateral institutional framework would dominate the political discourse.

Finally, by the beginning of the 1980s, Canada had eliminated approximately 80% of its tariff barriers, the bulk of which occurred as a result of the tariff cuts of the Kennedy and Tokyo Rounds (Lipsey 2000, 102). The trade liberalization that Canada had undertaken in the 1960s and 1970s spurred the private sector to rationalize production and move into higher value-added activities. By the early 1980s, more confident in their competitive abilities, Canadian industries began exploring opportunities to enhance commercial ties with their American counterparts in a tariff-free continent-wide market. In conclusion, after experimenting with trade diversification, Canada settled for economic policies that enhance market processes. This policy path became increasingly entrenched and reinforced by dynamics that resulted in liberal continentalism.

Chapter Five

The Limits of State Entrepreneurship and Gatekeeping

In elaborating on the purpose of the federal state's entrepreneurial and gatekeeper functions, Secretary of State for External Affairs, Mitchell Sharp, stated that it was the goal of the Trudeau government to encourage the "emergence of strong Canadian-controlled firms . . . by provid[ing] a climate conducive to the expansion of Canadian entrepreneurial activity [and that would seek particularly] to foster the development of large, efficient multinationally-operating Canadian firms that could effectively compete in world markets [and] meet a higher proportion of the domestic requirement for goods and services" (Sharp 1972, 11). Importantly, as industrial adjustment policy measures, the entrepreneurial and gatekeeper functions were not based on the idea that foreign capital was detrimental to the Canadian economy. Instead, through different courses of action, both functions were intended to augment the level of domestic control relative to that of foreign control and create incentives for foreign investors to contribute more in the development of Canada's manufacturing strength.

This chapter analyzes three policy measures that were designed to help increase domestic control of national economic activities. Established in 1971, the Canada Development Corporation (CDC), a state holding company operating as a public and private stock company, was given a mandate to enhance Canadian ownership and promote Canadian developmental ambitions by serving as a supplementary source of Canadian equity investment. Two years later, the Foreign Investment Agency Review (FIRA) was created to screen foreign investments in order to assess the scope of their contribution to the Canadian economy. Lastly, the National Energy Program of 1980 sought to Canadianize the domestic oil industry by augmenting the Canadian-controlled share of this industry to 50% by 1990. These three policy measures signified an important shift from the postwar position of accepting the high degree of foreign control as a cost of promoting sustained economic growth. It was believed that by playing the roles of entrepreneur and gatekeeper, Ottawa could help neutralize the control-growth tradeoff.

These three programs, however, never functioned the way they were originally conceived. This chapter will explain why that was the case. The FIRA, CDC, and NEP all represented federal policy thrusts that encountered strong provincial, societal, and bureaucratic opposition. This affected the way in which Ottawa proceeded with establishing those programs, particularly the organizational structure to support them. In the case of FIRA and the CDC, the political struggle in which policy entrepreneurs were embroiled ultimately led them to create two organizational structures that had limited capacities to enhance Canadian control of national economic assets. As for the NEP, the federal thrust into the energy sector provoked provincial riposte as the provinces, particularly Alberta, sought to assert their constitutional authority over this sector.

Canada Development Corporation

Although the idea of creating the Canada Development Corporation dates back to the 1957 Report of the Royal Commission on Canada's Economic Prospects (the Gordon Commission), it was not until the 1960s that the idea caught on in the government. Over the course of eight years, beginning in 1963, a public policy debate centering on two competing viewpoints emerged on the issue of the CDC. The debate ended in 1971 when Minister of Finance Edgar Benson introduced Bill C-219 in the House of Commons, the legislation that created the CDC.

From the outset, two schools of thought framed the public policy debate over the CDC. Developed by Walter Gordon, a prominent Liberal Party member who served as Finance minister under Prime Minister Lester Pearson, the nationalist school emphasized that the government needed to create an investment corporation whose chief goal would be to "buy back Canada." The Canada Development Corporation would assist Canadians in gaining greater leverage over domestic economic decisions in the face of high levels of foreign control of Canadian assets. As Gordon argued in 1965:

> The vast majority of Canadians want our country to remain free and independent. But if we lose our economic freedom—and an excessive absentee ownership of our businesses and resources means some loss of economic freedom—then sooner or later we shall lose our political freedom also. Many things must be done to prevent this. One of these is to secure in Canadian hands a greater measure of control of future economic developments. The Canada Development Corporation will help us to do this (quoted in Neufeld 1966, 30).

By providing industrial venture capital, the CDC was expected to help fix the capital gap, which, as some argued, was the principal reason for the high absentee ownership of Canadian industry. According to Gordon, the high presence of foreign capital reflected the lack of available domestic equity capital to finance developmental projects. The new investment corporation was expected to attract private capital and channel it into risky but potentially profitable developmental projects, while simultaneously working toward taking back Canadian industry. This approach was met by strong resistance from the business and financial communities in Canada. On the one hand, they rejected Gordon's claim that Canada faced a capital gap problem. As the authors of the Report of the Royal Commission on Banking and Finance noted: "We do not regard the sometimes heavy foreign borrowing by Canadian business in the postwar period as resulting primarily from deficiencies in the domestic financial markets" (quoted in Neufeld 1966, 31). On the other hand, the Canadian business and financial communities were hostile toward Gordon's approach because it would involve interfering with how the market makes investment decisions.

Gordon also had a difficult time persuading his political party to support his CDC plan. In his Throne Speech in 1963, Prime Minister Pearson stated that his government was committed to creating a corporation designed specifically to increase Canadian participation in domestic economic activities. Shortly after the budget speech, Gordon, who was the point man working the CDC file, submitted a resolution to the House of Commons calling for the creation of the CDC. Initially the resolution did not provoke opposition. However, in the wake of Gordon's 1963 budget speech in which he proposed a takeover tax as a way to make it more difficult for foreign companies to acquire Canadian companies, lawmakers began viewing the resolution as part of a nationalist design focused on thwarting foreign participation in the Canadian economy. As Liberal members attempted to shoot down Gordon's takeover tax proposal, the CDC resolution was caught in the line of fire and Gordon was forced to drop both plans.

The CDC issue was purposefully left out of Gordon's 1964 budget as well as Pearson's Throne Speech. As Kent (1988, 346) observes, "Action was inhibited in 1964 by the government's induced caution about anything that smacked of economic nationalism." Although the CDC plan resurfaced in 1965, both in Pearson's Throne speech and in Gordon's budget speech, no policy initiative emerged. However, if there was one thing that Gordon learned from his experience as being the lone voice of economic nationalism in Pearson's government, it was that the CDC would have to be equally concerned about making profits and basing its activities on sound business decisions. To that extent, the CDC would have to pursue commercial and nationalist objectives simultaneously.

The other school of thought, the developmental approach, was advanced by the new Finance Minister, Mitchell Sharp, when Gordon stepped down in November 1965. According to Sharp, the main objective of the CDC was to make sound investment decisions based on expected future earnings and opportunities for growth. According to Dimma (1974, 360), Sharp believed that the role of the CDC "would be chiefly to grow large and profitable through direct investment in viable Canadian enterprise and developments. Canadian economic independence would be promoted only to the extent that the CDC—which would be owned and controlled by Canadians—prospered through its role in industrial development and growth."

Thus, as the developmental approach held, the CDC would be assigned a commercial role, which, if pursued successfully, would enhance Canadian presence in the economy. Moreover, Sharp's approach maintained that it was important for the CDC to operate as independently of government involvement as possible—that is, function as a market-oriented commercial entity. The CDC's ability to mobilize private saving and to make sound investment decisions depended on the assurance of non-interference by the government.

When Gordon rejoined Pearson's cabinet in 1966, support was building for Sharp's developmental approach to the CDC. Sensing that there was now adequate attention being paid to the issue of Canadian control and foreign investment, Gordon persuaded Pearson to create a task force to study the effects of foreign investment in Canada. In its report *Foreign Ownership and the Structure of Canadian Industry* (Watkins Report), the Task Force on the Structure of Canadian Industry advocated Sharp's approach.

The Task Force's eight members—all of whom were economists—agreed that the CDC could be used "to enhance the capacity of the Canadian economy to grow with relatively less foreign direction investment" (Watkins Report 1968, 361). Specifically, the Task Force envisioned the CDC playing an active role in innovative projects in the areas of resource development as well as contributing to the rationalization of industry. These views were repeated in the Eleventh Report of the Standing Committee on External Affairs and National Defence (Wahn Report) completed in June 1970. The Wahn Report (1970, 33:94) noted the following:

> [The CDC's] main purpose should not be to buy back Canadian businesses now owned by Americans or other foreign citizens. Nevertheless, in exceptional circumstances it would be permitted to make investments to retain or establish a Canadian presence in a vital industry where it would serve as a pacemaker or a goad for foreign owned corporations. The Canada Development Corporation should not be . . . concerned with trying to outbid American businesses each time they try to buy Canadian concerns. Rather, the Canada Development Corporation . . . could arrange mergers among smaller Canadian companies which needed a large scale financial and administrative base and it might become an investor in such enterprises.

The two governmental reports envisioned the CDC as a holding company, into which Canadians and government invested, that would move into different industrial sectors for the purpose of contributing to the growth of the country's industrial strength and, in the long term, expanding the share of industrial assets under the control and ownership of Canadians.

When it was finally introduced by the Trudeau government in January 1971, Bill C-219 assigned two principal objectives to the CDC. As Section 2 of the Bill pointed out: "The purpose of this Act is to establish a corporation that will help develop and maintain strong Canadian controlled and managed corporations in the private sector of the economy and will give Canadians greater opportunities to invest and participate in the economic development of Canada." Central to the legislation was the commitment not to use the CDC in a way that impeded the inflow of foreign investment, but rather to ensure that the CDC was proactive enough to identify and secure profitable investment projects before foreign investors seized such opportunities.

Not long after the CDC Act was passed, the Globe and Mail (1971), which had echoed the opinions of the business and financial communities in previous years on the CDC issue, was less than enthusiastic about its creation. In an editorial titled "The Beast Is Loose," the newspaper cast doubts about the CDC's ability to balance its two mandates:

> What it comes down to is that the Government is trying to be half government and half entrepreneur. Neither one nor the other, only a maladroit mix, will set the cast of the entire corporation. Its motives will be forever suspect; its personality forever split; its decisiveness forever impaired. It will be afraid to take risks because of political repercussions and afraid to play safe because of business repercussions. Canadians may sympathize with its fervently proclaimed wish to protect Canadian economic independence. But they will fear its deformities and what those deformities can produce. Poor beast. It should be put to sleep rather than be unchained.

The Canada Development Corporation's Divided Mandate

As a mixed corporation, the CDC was expected to operate like a private enterprise and to mobilize private savings by promising strong returns on money invested in the holding company. To accomplish this, it was necessary for the CDC to have freedom from state control. To assure its relative autonomy, the CDC, unlike other Canadian Crown corporations, was not subject to the provisions of the Financial Administrative Act. Under this Act, Crown corporations were accountable to Parliament for their operation via the particular Minister responsible for their oversight. Because the CDC was exempted from Crown agency status, it did not have to seek budgetary approval from the Parliament nor meet financial reporting requirements, thus shielding the organization from parliamentary scrutiny and supervision.

The decision to lodge the CDC within the Department of Finance was also intended to give the organization the necessary flexibility to act in accordance with the profit motive. The Finance Department had a reputation for being the department least in favor of expanding state involvement in the economy and was deemed to be the best agent to ensure that the CDC upheld its commercial mandate with the least government interference. Had the CDC been placed in the Department of Industry, Trade, and Commerce or the Department of Regional Economic Expansion, there was the risk that the CDC would become implicated in the politics of industrial policy and of regional development to the detriment of profit maximization.

The CDC's investment decisions were expected to increase Canadian participation in the economy and diversify Canada's industrial base through a selective corporate strategy. When Bill C-219 was introduced, Benson noted that "able and experienced entrepreneurs will direct the corporation's operations to areas of critical importance in economic development—to high technology industry, to resource utilization, to northern-oriented companies and to industries where Canada has a special competitive advantage" (quoted in Brooks 1983, 529).

Several organizational features were fixed in place to ensure that its operations advanced these collective objectives with minimum hindrance to its profit motive. For example, in its first three years the CDC was completely state-owned, after which the corporation could begin engaging in public offering of shares. However, a 10% baseline was established that ensured that the government would always exercise some influence over the CDC. Moreover, such measures as the 3% limit on the share of stocks individual shareholders could own, the rule that all directors of the board had to be Canadian citizens and that the majority of them reside in Canada, and the requirement that all shareholders had to be Canadian residents or citizens, all helped make the CDC a policy instrument with which the government could advance the public interest.

The Performance of the Canada Development Corporation

In assessing the performance of the CDC, three questions should be taken into account: First, how well did the CDC fulfill its developmental objectives? Second, how well did the private corporate model upon which the CDC was based lure private capital? Finally, how well did the federal state restrain itself from interfering with the operations of the CDC?

The CDC did poorly in terms of diversifying its developmental activities. More than 80% of the CDC's investments went into the energy and mining sectors throughout the 1970s and early 1980s. Of importance were the petrochemical and oil and gas industries. The CDC's acquisition of all the shares of the state-owned Polymer Limited (later renamed Polysar) in 1972 and its investment in Petrosar starting in 1974 made petrochemicals the sector that received the most investment capital from the CDC until 1981. From 1981 to 1982, the oil and gas sector received the largest share of the CDC's investment capital, after it acquired Aquitaine Company of Canada Limited from its French parent company and the oil, gas, and sulfur assets of Texasgulf, both in 1981.

There were additional corporate moves that the CDC undertook in other industrial sectors. They include, for example, the CDC Data Systems' purchase of majority interests in AES Data Ltd., Wordplex Corporations, and Ventek Limited in mid-1978 and the CDC's acquisition of a 57% interest in the American photocopier maker Savin Corporation in 1982. These acquisitions made the information processing category the fourth largest sector to receive CDC investment. Still, the attention given to those industries paled in comparison to the CDC's interest in the energy sector.

Founded on the private corporate model, the CDC was designed to operate much like a private company and appeal to private investors by promising competitive returns. To achieve this goal, the CDC could begin issuing shares to the public three years after it was created, thus reducing the percentage of shares owned by government over a period of time. Although the government's holdings declined from 100% in 1974 to 68% in 1978 to 48.6% in 1981, the level of government ownership remained above the expected target. Disappointed at the lower than expected percentage of privately-owned shares, the government sought to encourage more private investors to channel funds into the CDC by "writing a letter of understanding to the Board of Directors and by refraining from exercising its right to vote in favor of the option to nominate 4 members to the 18-to-21 member board," as Laux and Molot (1988, 96) point out. Private investors were reluctant to invest in the CDC in part because they discounted the government's reassurance that it would not interfere in the corporation's affairs.

The other factor that contributed to low private sector involvement related to the lack of investor confidence in the CDC's ability to maintain a balance between its two mandates. A survey of shareholders' perceptions of the CDC taken in 1976 concluded that only 30% of respondents disagreed with the claim that the CDC was designed to buy back Canada from foreign investors (Kelly 1976). The perception among private investors was that the CDC was sacrificing profitability to pursue the nationalist mandate.

If private actors felt that the institutional devices intended to protect the CDC from government interference were weak, they were proven correct in the wake of the Liberal Party's return to power in 1980 after a brief Conservative interregnum. Emboldened by a renewed interest in the ideology of economic nationalism, the Liberal government unveiled an economic agenda that promised more government involvement in the economy. In particular, the new Minister of Industry, Trade, and Commerce, Herb Gray, suggested that the government exert more influence over the management of the CDC so that it could "take a stronger role in the manufacturing sector [that was] underrepresented in CDC's portfolio" (Whittington 1980, 2).

The willingness on the part of government to use the CDC as an instrument of policy was displayed in two cases. In late 1980, the Department of Industry,

Trade, and Commerce attempted to persuade the CDC to invest in Massey Ferguson, a Canadian multinational corporation, at a time when the corporation was seeking to refinance its operations. When the CDC opposed the plan to bail out the Canadian firm, the government pressured the Board of Directors to replace the current chairman, Frederick Sellers, with Maurice Strong whose views on the CDC were closer to the government's. Such interference soured relations with the CDC management, who strongly resisted this move by the government (Foster 1983).

Before a solution could be found to this impasse, the government further undercut the CDC with a plan to subsume the entity under a new holding company called the Canada Development Investment Corporation (CDIC). The economic recession that gripped Canada in the early 1980s, coupled with the revelation that an increased number of Crown corporations were facing financial and managerial problems, caused the government to look into ways of rationalizing publicly owned assets and to increase the profitability of commercially oriented public companies such as Teleglobe Canada, Eldorado Nuclear, and Canadair among others. Created in 1982 as an investment holding company with a Crown corporation status, the CDIC was given the mandate to bolster only the state-owned companies that remained economically viable and divest out of those no longer profitable.

The CDIC became the holding company for the government's 48% share of the CDC; for such Crown corporations as Canadair, De Havilland Aircraft, Eldorado Nuclear, and Teleglobe Canada; and for the government's $125 million equity in Massey-Ferguson (CDIC 1983). The CDIC was also designed to be a more tractable instrument of public policy. As Brooks (1983, 537) points out: "The origins of the CDIC cannot be considered apart from the inability of the government to exercise even a minimal level of control over the investment activities of the CDC. If the commercial success of the CDC was achieved . . . at the cost of its failure as a governing instrument, the reverse is a likely prognosis for the CDIC." To avoid a repetition of the political feud between the CDC and the government, an institutional choice was made to subject the CDIC to political accountability. In particular, the order-in-council (a cabinet decision) that created the CDIC established that "the Corporation shall comply with any specific direction in writing that may be given to it by the Governor-in-Council in furtherance of any policy or policies of the Government of Canada" (quoted in Laux and Molot 1988, 134).

The case of the CDC illustrates the difficulty Ottawa encountered when it sought to employ a well-established policy instrument, the creation of a public enterprise, to perform a task that had never been undertaken before. What distinguished the CDC from other public enterprises was the fact that it went beyond the normal scope of government involvement in the economy. Rather than the activity-specific functions of other Crown corporations, such as the involvement of Canadair in the transportation sector, Teleglobe Canada in the telecommunications sector, and de Havilland in the aerospace sector, the CDC was designed to move into as many sectors as possible. While undertaking this, the agency was required to balance two mandates that were not necessarily compatible. On the one hand, its commercial mandate forced the CDC to identify those sectors of the economy where there were opportunities for growth and profit making. Canada's abundant energy resources made this sector one of the very few in which the CDC could accomplish this objective. On the other hand, its nationalist mandate required the CDC to assist in expanding Canadian participation across all sectors of the economy, a task that demanded that the CDC diversify its activities into sectors that did not necessarily promise opportunities for sustained growth or profit making.

The effectiveness of the CDC in performing the gatekeeper function was also undermined as a result of the arm's-length pattern of government-societal relations, which constrained the government's ability to mobilize private savings. Whereas other Crown corporations did not depend on the consent of the private sector to attain their objectives, the CDC demanded it.

The Foreign Investment Review Agency

As a general rule, governments tend to be concerned with the level of foreign ownership and control of the productive assets of the national economy as well as with the performance of national firms abroad. Governments worry about foreign involvement because as the ratio of foreign to domestic ownership increases, the less the domestic interests are reflected in the economy's process of economic development. Moreover, a government takes particular interest in the international performance of its national firms because the economies of scale they obtain by expanding their international activities help boost national competitiveness and fuel economic growth. Historically, the Canadian state has paid attention to foreign participation in the economy; however, the issue gained particular importance starting in the late 1960s and early 1970s. Canada's heightened concern with foreign investment was illustrated when the Trudeau government created the Foreign Investment Review Agency (FIRA) in 1973.

Historical Antecedents

Since 1867, some sectors of the Canadian economy have been subject to foreign investment regulations. For instance, from the beginning of the 1900s to the late 1960s, a number of laws limited the level of foreign participation in the banking, transportation, and broadcasting industries. In the 1960s, the regulatory power of the federal state spread to the energy sector, as Ottawa enacted such laws as the Canada Mining Regulations, the Canada Oil and Gas Regulations, and the Northern Mineral Exploration Assistance Regulations—all of which were aimed at increasing the level of Canadian participation in a sector largely dominated by American companies (Franck and Gudgeon 1975, 96).

The passage of the Corporations and Labour Unions Returns Act (CALURA) in 1962 amounted to an expansion of the government's ability to monitor the activities of foreign companies. The Act required all Canadian companies with assets and revenues that exceeded a minimum level to file detailed annual reports revealing their operations, corporate assets, and composition of ownership structure. Finally, in response to the growing presence of foreign investors in Canada, Prime Minister Pearson's Liberal government in 1966 introduced "Guidelines of Good Corporate Citizenship," a set of non-compulsory rules intended to influence the practices of foreign corporations. In assembling a list of guidelines, Trade Minister Robert Winters hoped that "this statement of the basic essentials of good corporate citizenship will contribute to a better understanding of the role of subsidiaries . . . in our economy and will encourage and facilitate their full participation in our growth and development in line with Canada's trade, economic, and social needs" (Godfrey and Watkins 1970, 52).

In the few years before the creation of the FIRA, three important governmental reports that studied the issue of the political and economic effects of foreign investment in Canada were published. These reports were *Foreign Ownership and the Structure of Canadian Industry* (Watkins Report, 1968), the *Report of the Standing Committee on External Affairs and National Defence* (Wahn Report, 1970), and *Foreign Direct Investment in Canada* (Gray Report, 1972). The most important conclusion of the Watkins Report was that the practice of extraterritoriality—where foreign corporations abide by the laws and policies of their home governments in the host country—was the most onerous outcome of hosting a large number of foreign firms. As the report (1968, 49) noted: "From the viewpoint of the host country, the extraterritorial extension of law is an undeniable cost of foreign direct investment. Overlapping legal jurisdiction threatens the national sovereignty of the host country [and] if the content of the law conflicts with the policy of the host country, the situation simply becomes more tense."

The Watkins Report also pointed out that the threat of extraterritoriality would increase as the level of foreign ownership and control increased. To enhance the benefits derived from the activities of multinational corporations operating in Canada without undermining Canada's reputation as a lucrative investment location, the Watkins Report (1968, 395-6) recommended the creation of a special agency that would survey and monitor the activities of multinational corporations in Canada and seek to gain their commitment to fulfilling the 1966 Guidelines.

The report submitted by the House of Commons Standing Committee on External Affairs and National Defence (Wahn Report) recorded the first time members of Parliament had addressed the implications of foreign investment in Canada. Unique in this respect, it was also unique in the recommendation it put forth:

> The Committee recognizes that as a general rule it is desirable that Canadians should control Canadian companies by owning at least 51% of their voting shares, particularly in the important sectors of the economy where American control is now most highly concentrated, and that we should move toward this goal as rapidly as capital requirements and other relevant circumstances permit.

According to the Wahn Committee, the current level of foreign investment in Canada had reached a level high enough to warrant such a radical buy back policy. Interestingly, the committee did not consider how much of an expansion of state intervention would be required to achieve such a vast repatriation of foreign-owned economic assets. However, the committee did recommend that the government create a Canadian Ownership and Control Bureau tasked to screen foreign takeovers of Canadian business, scrutinize all applications for new foreign investment in Canada, and assess the activities of those multinational corporations where the practice of extraterritoriality was alleged.

The period from 1970 to 1973 also witnessed some important changes that helped keep the issue of foreign investment regulation at the forefront of the government agenda. Public perception of foreign investment shifted in favor of foreign regulation, as opinion polls indicated. When Canadians were asked in 1970 whether there was enough American investment in Canada or whether they would like to see more, 62% of them stated that there was enough, while 25% wanted to see more investment. In 1975, 71% of Canadians noted that there was enough American investment in Canada, while 16% wanted to see more. Equally significant were the responses to the question of whether Canadians would support a policy initiative to buy back a majority control of American companies in Canada even if that meant having to accept a reduction in their standard of living. In 1970, 46% of Canadians supported a policy that promoted the repatriation of economic assets, while 32% disapproved. In 1975, the approval rate had increased to 58%, while 26% rejected such an economic plan (Molot and Williams 1984, 93).

This period also saw the emergence of two influential societal groups that advocated stricter regulations on American investment in Canada. Founded in 1969, the Waffle entered the political scene by forming an alliance with the New Democratic Party (NDP). The position of the Waffle was that "the major threat to Canadian survival [was] American control of the Canadian economy." The most important issue facing Canadians, as they saw it, "was not national unity but national survival, and the fundamental threat [was] external not internal" (quoted in Brodie and Jenson 1988, 273). To counter this external threat, the Waffle and the NDP proposed a plan to drastically increase the role of the state in the economy to the point of nationalizing various sectors of the economy.

As a centrist organization, the Committee for an Independent Canada (CIC) was created in September 1970 to offer a more moderate alternative to the radicalism of the Waffle. The CIC urged Ottawa to "adopt legislative policies that [would] significantly diminish the influence . . . exerted by outside powers on Canadian life." Although it claimed to have non-partisan ties, the CIC maintained ties to the Liberal Party. The former Liberal finance minister Walter Gordon served as its honorary chairman and Liberal Party members such as Herb Gray and Alastair Gillespie, both of whom were instrumental in defining the statist variant of the Third National Policy, were supporters of the CIC. Thus, on the eve of the publication of the Gray Report, a wave of economic nationalism had swept throughout Canada, spurred by an emboldened and more dominant middle class (Azzi 1999).

When the Gray Report came out in 1972, it was the most comprehensive analysis of foreign investment in Canada ever undertaken by a government-sponsored commission. Unlike the Watkins Report and the Wahn Report, the Cabinet had called on Herb Gray in 1970 to bring "forward proposals on foreign investment policy for the consideration of the government" (Gray Report 1972, v). Although the Report included a caveat that "it [was] not a statement of government policy nor should it be assumed that the government endorses all aspects of the analysis contained in it," the report had a defining impact on subsequent public debates on the issue of foreign investment restrictions.

The Gray Report put forth two groundbreaking conclusions. First, it concluded that the effects of foreign-controlled Canadian assets went beyond extraterritoriality. While acknowledging the economic gains associated with foreign investment, the Gray Report contended that the high level of foreign penetration contributed to the "truncation" of Canadian manufacturing. The Report (1972, 42-3) defined the problem as follows:

> Truncation means potentially less decision making and activity in Canada— fewer opportunities, fewer supporting services, less training of local personnel in various skills, less specialized product development aimed at Canadian needs or tastes and less spill-over economic activity. It ties the subsidiary to the parent in a relationship of dependence.

In light of this, the report proposed the establishment of an agency to evaluate the contributions of foreign investment on the basis of what "significant benefits" they conferred on the Canadian economy.

The second conclusion of the Gray Report was that the government needed to enhance its influence over foreign investors. The report considered three approaches to achieve this. Whereas the "key sector" approach proposed restricting foreign involvement in certain "commanding heights" of the economy, the "fixed rules" approach suggested setting a mandatory minimum level of Canadian equity ownership in American-owned companies. Rejecting these two approaches, the Gray Report (1972, 462-9) advocated the investment screening approach. It envisioned a screening agency with the capacity to scrutinize all foreign takeovers of companies operating in Canada, all new foreign investment entering Canada, and all expansions into unrelated activities of foreign-owned companies already established in Canada.

The Politics of the FIRA

The politics surrounding the creation of the institutional machinery of the FIRA were profoundly shaped by provincial, bureaucratic, and societal forces. The nature of the public policy issue—restricting foreign investment—inevitably involved the provinces because of jurisdictional overlapping. To make matters more complicated, the Liberal government of Trudeau was as divided over foreign investment regulation as was the federal bureaucracy. Moreover, the political battle took on added significance because the proposal to restrict foreign investment would mean expanding the interventionist capacity of the federal state beyond any level previously experienced.

In early May 1972, the Trudeau government introduced the Foreign Takeovers Review Act (Bill C-201) in the House of Commons. The Takeovers bill did not adopt in full the recommendations that the Gray Report had made. In fact, Bill C-201 proposed to review only foreign takeovers of existing Canadian businesses, thus leaving unscreened all existing foreign investments and all new foreign investment projects in Canada.

The reluctance of the Trudeau government to broaden the scope of foreign investment screening was owed to three factors. First, policymakers faced a steep learning curve, knowing little about how to deal with the administrative and managerial complexity of engaging in foreign investment review. Second, governmental and societal support for an open foreign investment regime remained strong in the early 1970s. Finally, foreign investment was viewed differently across the provinces, for foreign investment tended to serve the interests of some provinces better than others. Thus, while there was a growing awareness of the drawbacks associated with the elevated level of foreign presence in Canada, there was only enough support to create a mild screening mechanism (Franck and Gudgeon 1975, 105-6).

The lifespan of Bill C-201 was cut short, however, when Parliament was dissolved in September 1972 in preparation for the October election. After a couple of months out of the political limelight, the foreign investment review issue reemerged when the minority government of Trudeau introduced the Foreign Investment Review Act (Bill C-132) in late January 1973.

Two key factors helped push the new bill through the legislative process. It is without a doubt that the new electoral landscape had weighed heavily in shaping the Review Act. To appease the New Democratic Party, which became the pivotal party in the House of Commons, the Trudeau government had included a few of the recommendations that the NDP had requested when Bill C-201 was introduced. The other factor was the appointment of Alastair Gillespie as Minister of Industry, Trade and Commerce. As an advocate of economic nationalism and influential political figure, Gillespie was responsible for shepherding the bill through the policymaking process. His commitment to the piece of legislation led the bill to receive Royal Assent in December 1973.

The FIRA received poor marks during the twelve years if its existence. According to Molot and Williams (1984, 89): "The agency was more an exercise in symbolic politics than a genuine effort to regulate foreign investment coming into Canada." And, as Franck and Gudgeon (1975, 109) note, "It is evident that the Government consistently sought to strike a balance between prohibition or roll-back of foreign investment on the one hand and traditional laissez-faire on the other." These views are well reflected in the agency's track record. The percentage of rejected applications gradually declined over the course of the 1970s, dropping from 18% in 1974 to 5.01% in 1977 (Jenkins 1992, 116). What explains the FIRA's poor marks? The answer is two-fold: it is due in part to what it was programmed to do and in part to the organizational structure the agency was given.

One of the shortcomings of the FIRA was that its scope of authority was restricted and ambiguously defined. The FIRA was given the authority to review three types of foreign investment: first, most acquisitions of Canadian corporations by non-Canadian investors; second, the establishment of new Canadian corporations by foreigners who were not already present in Canada; and third, new investment into an unrelated line of business by non-Canadians already operating in Canada.

Omitted from the FIRA's scope of authority was the ability to review the operations of existing foreign companies in Canada, particularly their expansion and investment into related production activities. This was an important omission because this type of investment activity represented 80% of all foreign investment activities in Canada. In a brief submitted to the House of Common's Standing Committee on Finance, Trade and Economic Affairs, the Committee for an Independent Canada noted that the FIRA "assumes that takeovers and new foreign investment may or may not be of significant benefit to Canada, but it refuses to apply this same assumption to existing foreign enterprises which represent the bulk of our problem of controlling the Canadian economy in the interest of Canadians. It establishes a 'grandfather' clause to legitimize all existing foreign investment, whether good or bad."

The FIRA's authority was further undercut because the criteria used to review foreign investment were ambiguously defined. The FIRA assessed foreign investment on the basis of whether it contributed "significant benefits" to Canada. But, without specific benchmarks, there was no way to measure the significance of foreign investment contribution. In justifying the decision to go forward with a general set of criteria, ITC Minister Gillespie stated during a committee hearing on Bill C-132 that "at this stage, precise standards for measuring acceptability cannot be spelled out. Ability to spell these out will depend upon experience with specific cases; particular decisions will lead to a body of guidelines [that] I would hope then, it may eventually be possible to publish." In the meantime, five general criteria were established:

(a) The effect on the level and nature of economic activity in Canada, including the effect on employment, on resource processing, on the utilization of Canadian parts, components and services, and on exports;
(b) The degree and significance of participation by Canadians in the business enterprise and in the industry sector to which the enterprise belongs;
(c) The effect on productivity, industrial efficiency, technological development, innovation and product variety;
(d) The effect on competition within any industry or industries in Canada;
(e) Compatibility with national industrial and economic policies (FIRA 1976, 11).

Two factors contributed to the problem of ambiguous guidelines. As Franck and Gudgeon point out, one of the sources was that the government was torn between letting market players operate relatively unimpeded as they made investment decisions or regulating their behavior. The other source of the problem had to do with the divergence of interests among the provinces regarding the foreign investment regulation. Inter-provincial differences were clearly expressed when the provincial governments debated Bill C-132. From the viewpoint of the Ontario government, the main purpose of the FIRA would be to contribute to ongoing efforts to create a more integrated national economy, which meant not interfering with the hinterland-center pattern of economic development. Although Ontario recognized that it was necessary for the FIRA to "be sensitive to diverse provincial viewpoints," the province argued that "the wider interventionist role which will be available to the federal government should not be used as an instrument of regional development policy." Clearly, the position of the Ontario government was that the FIRA's screening capacity was not to interfere with the prevailing regional distribution of foreign investment—that is, persuade foreigners to invest in another province when the original plan was to invest in Ontario.

The government of Saskatchewan had a different position than Ontario, for it suggested that the FIRA should attempt to alleviate regional disparities caused by regionalized economic specialization. As Saskatchewan's Minister of Finance Elwood Cowley pointed out, one of the consequences of the postwar open foreign investment regime was that "foreign ownership increase[d] regional disparities and thus hurt a province like Saskatchewan." Foreign investors, according to the Minister, reinforced the hinterland-center pattern of economic development by locating the bulk of their operations in central Canada. Therefore, the FIRA's screening powers should be employed in a manner that sought to "ensure that investment is allocated across Canada by province and region such that the total package will contribute to a lessening of regional disparities."

But perhaps the harshest criticism leveled against the FIRA came from the Atlantic provinces. In conveying the region's position on the regulation of foreign investment, Stephen Weymen, President of the Atlantic Provinces Economic Council, remarked: "It seems as if certain segments of opinion in

Central Canada, having achieved the benefits of industrialization for themselves, now wish the Atlantic Provinces to forego the industrialization they never had in the interests of maintaining an ill-defined Canadian economic independence." Sharing the same sentiment as Mr. Weymen, the Premier of New Brunswick, Richard Hatfield, remarked:

> The Government of New Brunswick's opinion on the proposed legislation [to create the FIRA] is a simple and straight-forward one—we see no need for it in our region of Canada and we believe that it would prove to be harmful to our attempts to attract and to accelerate industrial development. The Maritime region has enough difficulty now in attracting industrial development. We are dependent to a large degree on foreign-controlled enterprises in our efforts to attract new industry. . . . This Bill will tend to have a disproportionate impact on an area like New Brunswick because foreign-controlled firms are often the only one which can be interested in our area.

In light of these divergent provincial views and the ideological division within the federal government, it is no wonder that efforts to fashion a set of specific guidelines by which to evaluate foreign investment failed.

The Organizational Limits of the FIRA

The proponents who expected the FIRA to be equipped with formidable organizational wherewithal were disappointed by what Bill C-132 had in store for the new agency. The legislation created an agency in the image of a non-departmental, advisory body. As such, the FIRA's stature would forever be diminished. Several factors contributed to this organizational decision. First, when the Gray Report recommended the establishment of a review agency, it argued against endowing the agency with independent decision-making capacity. According to the report (1972, 480), granting the agency an independent tribunal status "would require the government to delegate a substantial measure of responsibility for decisions which would have great importance to its overall industrial and commercial policies to an independent body." Accordingly, the government opted to give the agency an advisory function. The FIRA was tasked to process applications using the "significant benefits" standard. Its decision was reviewed by the Minister of Industry, Trade and Commerce, who offered his recommendation to the Cabinet. The Cabinet had the last say, either approving or disapproving the Minister's recommendation.

The decision to give the FIRA a non-departmental status reflected the political dynamic within the executive administrative apparatus. For one thing, no department offered the FIRA an appropriate home within which to operate. As Schultz, Swedlove, and Swinton (1980, 19) point out, the most likely "departmental candidates for administering the screening process were rejected as being too biased in favour of foreign investment." The Department of

Finance, for example, contended that restricting foreign investment would produce balance of payments problems. As for the Department of Industry, Trade and Commerce (ITC), the other departmental candidate, its bureaucrats were of the view that foreign investment played a crucial role in Canada's efforts to strengthen its industrial base. As Williams (1983, 166) observes, the policy ideas that the FIRA was expected to apply were "just too unpalatable to ingest for many senior bureaucrats" in both departments.

Another factor that contributed to the FIRA's non-departmental status was Trudeau's rational management reforms that were designed to centralize decision-making authority within the executive administrative apparatus. It was argued that the non-departmental status of the FIRA would give central agencies greater access to this new entity than if it were lodged within a department. As Schultz, Swedlove, and Swinton (1980, 20) note: "A non-departmental status for the agency, especially given its limited size and its concomitant inability to call on the resources of a 'parent' department would be a means whereby central agencies could attempt to ensure that they would maintain control of the process."

Finally, the FIRA fell victim to the influence of political expediency. Its organizational structure was ideal for the government to the extent that it could take credit when the FIRA received positive endorsements, and dodge criticism when it made controversial decisions concerning the status of applications. In sum, the FIRA's poor marks reflect the fact that its organizational structure afforded the agency little autonomy of decision and that it had no capacity to assert itself.

A Second Try at Gatekeeping and Entrepreneurship

In 1980, Ottawa announced two economic policy changes. The first policy change called for 'beefing up' the FIRA. In his Throne Speech that year, Trudeau stated that "my ministers believe that the stake of Canadians in their own economic destiny must be strengthened." Accordingly, "[t]he Foreign Investment Review Act will be amended to provide for performance reviews of how large foreign firms are meeting the test of bringing substantial benefits to Canada [and] amendments will be introduced to ensure that major acquisition proposals by foreign companies will be publicized prior to a government decision on their acceptability." The intent of this latter measure was to create a first-move advantage for potential Canadian investors. By publicizing foreign investment proposals, Canadian investors would be made aware of an acquisition opportunity that they may not have known about otherwise. Furthermore, Ottawa promised to "assist Canadian business wishing to repatriate assets or to bid for ownership or control of companies subject to takeover offers by non-Canadians."

The other major policy change was the creation of the National Energy Program (NEP) in 1980. The NEP had three policy objectives. First, it sought to promote Canadian self-sufficiency in energy supply by speeding up the exploration and development of high-risk sources of energy in Canada, such as those located in the Arctic, the offshore region of the Maritime provinces, and Western Canada where there was an abundance of heavy oils and tar sands. Second, the NEP sought to promote Canadianization by increasing the presence of state-owned corporations and Canadian-controlled firms so as to be able to control 50% of the oil and gas industry by 1990.

To achieve this, three instruments were to be used. First, the government would help lower, for Canadians, the cost of entry into the oil and gas industry, which was currently dominated by foreign concerns, mostly American. Moreover, the goal of increasing Canadian involvement in this industry was to be promoted by requiring that a 25% buy-in option be given to the federal government on any leases on federal-owned Canadian lands. The 25% buy-in option was designed to give Petro-Canada, a federal Crown corporation, an opportunity to increase its presence if it chose the option. Lastly, the FIRA's screening authority would be marshaled to promote the government's Canadianization goals.

The third objective of the NEP was to distribute rising oil and gas revenues in an equitable way. As Pratt (1982, 55) notes: "increased share of rents skimmed from the development of existing Western Canadian oil and gas reserves [were to be] appropriated by energy users, via low prices, and the federal government, via new taxes, while the share of rents accruing to producers and the producing provinces [was to be] substantially cut." One of the outcomes of this distributive plan was that the domestic price of oil was 3% below that of the United States, 108% below Great Britain's, and 137% below Italy's (Energy, Mines and Resources Canada 1982, 90). The new taxes on oil and gas enabled the federal government to increase its share of revenues from 12% in 1980 to 27% in 1982, while provincial shares dropped from 42% to 32% (Milne 1986, 85).

The NEP came under heavy criticism from many provinces. Leading the pack was Alberta, Canada's largest energy producing province. Since 1930, when the Western provinces were granted the right to own the public lands located within their confines, this region of Canada had greatly benefited from the rising world demand for oil. When Joe Clark's Conservative government came into power in 1979, he proposed a set of energy policies that promised to improve, in particular, the energy revenue of Alberta. Clark planned gradually to increase the domestic price of oil to about 85% of world prices, to privatize Petro-Canada, and to increase the federal government's tax base by seeking a revenue-sharing regime to which the provinces could agree (Pratt 1982, 33-4). However, Ontario, Canada's largest energy consuming province, staunchly opposed Clark's energy policy, for it meant that Ontario's economy would have to shoulder higher energy costs while Alberta enriched itself.

The intergovernmental dynamics changed when the Liberal Party returned to power in 1980. Ontario and New Brunswick sided with Ottawa's energy program and the other eight provinces formed an alliance opposing the NEP. The fundamental cause of provincial opposition to NEP was that the program amounted to a direct intrusion into provincial jurisdiction and threatened to undermine the provinces' capacity to pursue their own economic development strategies.

Ottawa was forced to amend the NEP in 1981 after its relations with Alberta soured into bitter confrontation. In May 1980, Alberta's Energy Minister, Merv Leitch, proposed a bill that would authorize Alberta to reduce oil supplies to the rest of Canada. In response, Ottawa proposed to impose a tax on exports of natural gas, a measure that would particularly injure Alberta since it was a leading natural gas exporting province. This pattern of thrust and riposte came to an end in September 1981 when Ottawa and Alberta reached a mutually beneficial five-year agreement in which both agreed to a schedule of energy price increases and a new revenue-sharing regime. In the following months, intergovernmental tensions were further defused as Ottawa entered into similar arrangements with British Colombia and Saskatchewan. In the end, these intergovernmental arrangements significantly softened the centralist and unilateralist aspects of the NEP.

Opposition to Ottawa's aggressive, unilateral policy changes also came from the Business Council on National Issues (BCNI). The business group was concerned that the United States would counter the activities of the FIRA by employing similar discriminatory measures against Canadian investments in the United States. At a time when Canadian firms were investing more and more in the United States, the specter of American foreign investment regulations led the Canadian business community to pressure its own government to eliminate foreign investment regulations and establish a level playing field in which foreign investors would be subject to the same treatment as national firms. On the other hand, the BCNI was worried that a tougher FIRA would undermine Canada's reputation as a favorable investment location. Thus, the BCNI demanded that "the Canadianization rules be relaxed, and that curbs on foreign-owned companies by softened" (Clarkson 1982, 81).

Ottawa's nationalist inspired policies also drew fire from the American government. In the 1970s, American officials tolerated the FIRA. In explaining Washington's position on the agency during the decade, American Ambassador to Canada, Thomas Enders, observed: "It was feared that FIRA might act as a barrier to new incoming investment in Canada. But rather it . . . applied its mandate—to assure benefit to Canada in investment proposals. I can understand how Canada relying as heavily as it does on outside investment feels the need for having such a mechanism to insure that its interests are identified and met" (quoted in Clarkson 1982, 87). In fact, throughout the 1970s, the U.S. government directed most of its criticism on the foreign investment regulations of other countries because they were more arbitrary and discriminatory than Canada's.

All of this changed when Ottawa announced its plan to toughen the FIRA. Fundamental to the emerging Canada-United States discord was that, as Washington was seeking to liberalize economic exchanges with its allies abroad, Canada was moving in the opposite direction by introducing greater regulations. As Raymond Waldmann, then Assistant Secretary of the Finance Department for International Economic Policy, put it, "Canada's growing reliance on government intervention to direct and control national trade and investment patterns conflicts with this Administration's advocacy of free market oriented policies [and] pose[s] a threat to the international trading system." In response, Washington considered a number of retaliatory measures. First, the American Congress—having held several hearings on the FIRA—proposed two bills that would have imposed tougher standards on Canadian investment in the United States. Second, the Reagan administration considered using the Section 301 clause of the Trade Act of 1974—a retaliatory instrument used against countries that restrict American commerce. Finally, the American government requested that the General Agreement on Tariffs and Trade (GATT) examine the legality of the FIRA. With pressure coming from Washington, the business community, and the provinces, the Trudeau government was forced to throw out its plan to toughen the FIRA.

Washington was less effective at pressuring Ottawa to back away from the NEP. The fact that the NEP enjoyed strong public support helped embolden the Trudeau government to resist American criticism. But the greatest impediment facing the American government as it sought to persuade the Trudeau government to soften the nationalist edges of the NEP was the political deflection caused by the jurisdictional disputes between the provinces and Ottawa. The battle between Western Canada and Ottawa consumed most of the latter's attention. Furthermore, the American government could not count on Alberta and other provinces to fight Ottawa on its behalf. When Ottawa and the energy producing provinces found a settlement to the jurisdictional dispute, the bilateral arrangements preserved the Canadianization aspect of the original NEP. This was an outcome that fell short of what the United States desired (Clarkson 1982).

By the end of 1981, it became clear that the implementation of Ottawa's 1980 policy agenda had stopped in its tracks. Talks of enhancing the stringency of the FIRA abruptly ended after the Canadian government acknowledged in its 1981 budget paper that the existing apparatus was sufficient to accomplish the government's goals. It also added that "the special measures being employed to achieve more Canadian ownership and control of the oil and gas industry [were] not appropriate for other sectors," thus mollifying concerns about the "NEP-ing" of other sectors in the Canadian economy. In the years to come, the FIRA was gradually liberalized and was no longer used to support the NEP's Canadianization goal.

The Mega-Project Episode: A Different Kind of State Entrepreneurship

When Prime Minister Trudeau announced in his 1980 Speech from the Throne that the Liberal government would "promote a national development policy," the Cabinet was grappling with two competing development strategies. One strategy, advanced by Minister of Industry, Trade and Commerce Herb Gray, called for the allocation of $2.75 billion over the next four years to programs aimed at reducing Canada's reliance on foreign technologies, strengthening Canada's manufacturing and processing capacity, and increasing exports of higher value-added products. The Gray proposal, which was presented in early July 1980 as a Cabinet discussion paper, was based on the claim that the branch-plant activities of foreign-owned Canadian subsidiaries failed to contribute to Canada's efforts to expand its domestic technological base—which was imperative for moving into more sophisticated forms of manufacturing. In a speech titled "Economic Nationalism and Industrial Strategies," Gray (1980) noted that:

> The very high level of foreign involvement in our economy has serious adverse effects for the country's ability to fulfill national and regional development goals. [These effects include] inefficient, low-volume scale of many Canadian manufacturers, a low-level of R&D expenditure, a high propensity to import manufactured goods, difficulties associated with the export of manufacturers, and an inability to exploit for our advantage the potential linkages between natural resource wealth and the development of a healthy secondary manufacturing sector.

Gray's strategy sought to promote, through state incentives, such high-technology industries as aerospace, transportation equipment, and electronics, and to increase the state's export-financing efforts to boost exports of such goods. It also offered assistance to Canadian firms seeking to repatriate foreign-owned assets, and recommended that foreign investment regulations be toughened to compel branch-plant subsidiaries to contribute more to Canada's developmental endeavors. The strategy proposed to empower the Canada Development Corporation to assume a more active role in the manufacturing sector and enact an 'industrial benefits' legislation that would enable Canadian companies to serve as suppliers and subcontractors in major energy projects worth more than $100 million. Finally, the strategy called for the rationalization of troubled industries such as clothing, textiles, and footwear; a measure that would facilitate the transfer of capital from uncompetitive sectors to those manufacturing sectors with better prospects for growth (Whittington 1980, 1-2).

The second strategy was articulated in the Medium Term Track (MTT) document produced by the Ministry of State for Economic Development (MSED)—a central agency created in late 1978. Circulated in May 1981, the

MTT document stressed the importance of making full use of Canada's comparative advantage in resource production as a means of developing backward and forward industrial linkages. As the document pointed out,

> A fundamental and essentially permanent shift has taken place during the 1970s, one that strengthens Canada's traditional comparative advantage in the production of basic commodities [and] related manufacturing products . . . on the one hand, and increases the comparative disadvantage of many standard manufactured end products on the other. The shifts in the terms of trade toward basic materials can be expected at least to be maintained in the medium term, and perhaps even to strengthen (quoted in Doern 1983, 229-30).

This policy had two advantages over Gray's proposal. First, its strategic focus was broader in that it paid attention to the development of both the resource processing and manufacturing sides of the Canadian economy. Second, by focusing on both forms of production activities, the MTT document's design was sensitive to the regional division of economic activities.

The MSED strategy received an important endorsement with the release of the Blair-Carr Task Force's Report, *Major Canadian Projects: Major Canadian Opportunities*, in June 1981. Created in November 1978 by the Federal-Provincial Ministers of Industry Conference, the task force was a business-labor consultative body of approximately 80 senior Canadian business and labor representatives, co-chaired by Robert Blair, president of Nova, and Shirley Carr, executive vice-president of the Canadian Labor Congress. The report put together a list of mega-projects with an estimated total value of about $400 billion by the year 2000. These mega-projects were major resource development projects, each with an initial capital investment cost exceeding $100 million, and intended to have positive effects on the Canadian technological and industrial base.

As the MSED had envisioned in its plan, the mega-projects proposal was designed to create synergies between resource and industrial development, primarily by capitalizing on Canada's comparative advantage in energy resources. The Blair-Carr Report estimated that about 87% of the $400 billion spent on major projects would be in such sectors as electric power generation and transmission, conventional hydrocarbon exploration and development, heavy oil development, pipelines, and hydrocarbon processing and petrochemicals. Moreover, in terms of the geographical distribution of these major projects expenditures, the Atlantic region was expected to receive a little over 10%, central Canada 39%, western Canada 25%, and British Columbia, Yukon, and the Northwest Territories 25%.

The report indicated that the benefits of pursuing mega-projects would be manifested throughout the economy. In terms of opportunities for the labor market, mega-projects were expected to create a demand in high skilled occupations such as draftsmen, scientists, technicians, engineers, and managers. In terms of boosting technology-intensive manufacturing, the report (1981, 43) noted that "about one half of the investment in energy projects [would go to] manufactured products [such] as basic steel, pressure vessels, mining equipment, electrical cable and switch gear, offshore platforms and support vessels, instrumentation, and power generation and transmission," just to name a few.

The government responded to the Report by establishing in August 1981 the Office of Industrial and Regional Benefits within the DITC. Among the objectives pursued by the Office were, first, to encourage the kind of sourcing of equipment and services in Canada that offers a substantial degree of technology innovation domestically; second, to foster research and development activity in Canada and encourage world product mandating by foreign-owned firms in Canada; and third, to ensure maximum participation by Canadian companies and workers in all aspects of major projects development (Gray 1981, 9).

The *Statement on Economic Development for Canada in the 1980s*, a white paper tabled with Finance Minister Allan MacEachen's November budget, adopted the approach advanced by the MSED and the Blair-Carr Task Force. The principal argument of the white paper was the following:

> The leading opportunity [for the Canadian economy] lies in the development of Canada's rich bounty of natural resources. There is increasing world demand for Canada's major resources, such as petrochemicals, and further expansion of agriculture, forest-based industries and mining. Linked to this growth dynamic is manufacturing activity, both to supply machinery, equipment and materials needed for resource development and to extend the further processing of resource products beyond the primary stage. The third direction of growth linked to resources [is] built upon our technological infrastructure. This class of opportunities includes adaptations of Canadian and foreign technology and the provision of high technology services such as project management, engineering, industrial design and computer applications.

The government's decision to adopt the mega-project strategy reveals how Ottawa's entrepreneurial role was changing in the early 1980s. First, the government decided to let the private sector (foreign and domestic firms) be at the forefront of major project development, settling the debate as to how Ottawa would promote the mega-project strategy. When the Blair-Carr Task Force deliberated on the issue of state participation, labor and business representatives were at odds with each other. Labor representatives favored using public ownership and direct grant assistance, whereas business representatives preferred tax expenditures such as tax credits and accelerated write-offs. The government's decision to encourage private sector involvement via tax incentives and other market instruments signified that Ottawa was no longer willing to use public ownership or other interventionist instruments to bring about industrial adjustment. This move coincided with Ottawa's announcement that it would neither toughen the FIRA nor extend the Canadianization component of the National Energy Program to other sectors of the economy.

Second, the federal government had no desire to redefine Canada's comparative advantage via the heavy governmental hand. Instead, Ottawa's emerging economic adjustment plan placed a premium on the development of the country's vast natural resources and its technological spin-off effects. However, by depending on this economic sector, the economy became more vulnerable to volatilities in international market conditions. Indeed, that is precisely what occurred within months of announcing the mega-project development plan. The sharp decline in world oil prices and the lowered projections of future oil prices rendered some mega-projects unfeasible and postponed others.

Finally, the new development policy offered a solution to the regional bias that past federal economic policies often had. By exploiting Canada's comparative advantage in resources, while simultaneously developing backward and forward linkages, benefits would flow to both central and peripheral regions of Canada. As Leslie (1987, 20) points out, the new federal economic development policy was a "strategy of regional complementarity" because the "development of centre and periphery would be mutually reinforcing." Thus, by integrating regional and sectoral development, Ottawa's new entrepreneurial role could help defuse the intergovernmental conflicts stoked by previous regionally-biased federal actions.

Chapter Six

The Market as a Political Economic Solution

Liberal continentalism rose to prominence during the 1980s because of a convergence of particular systemic conditions and domestic forces. Central to the new adjustment strategy was the notion that free trade with the United States would generate domestic incentives for the federal and provincial governments to substitute market-enforcing regulations for political discretion as the principal guiding force to bring about industrial adjustment.

In the last two chapters, I have demonstrated how institutional dynamics prevented government officials from pursuing successfully interventionist policy measures that were nationalist inspired. Moreover, I have shown that Canada's political-economic institutional system has been more amenable to supporting policy solutions that enhance market processes. In this chapter, I will explain why Ottawa was able to build a coherent policy regime around liberal continentalism—a policy regime that has been stabilized by self-reinforcing mechanisms since the mid-1980s. Part of the explanation is based on international changes. Canada's rising level of vulnerability dependence vis-à-vis the United States—a product of increasing cross-border production integration and investment in relation-specific assets—and its modest improvements in its relative position led Ottawa to adopt the logic of integration. Informed by this logic, Ottawa sought to institutionalize the bilateral economic relationship—that is, anchor it to common rules and dispute-settlement procedures—as a way to minimize the costs of vulnerability dependence and create opportunities for economic growth.

At the domestic level, two factors contributed to the rising support for the expansion of market governance. On the one hand, the economic slowdown in the early 1980s coupled with a decade of failed interventionist policies rendered the current economic policy regime untenable. Government officials and the business community believed that liberal market reforms—such as privatization, deregulation, and free trade—would help bring about economic recovery and spur the kind of industrial rationalization and productivity gains that had evaded the strategy of economic nationalism. Liberal continentalism also served a political purpose. After a decade of intergovernmental tensions—partly fueled by Ottawa's economic nationalist policies—there was a need to find a set of policies that would help defuse such conflicts. Free trade and domestic market reforms provided a way for provincial and federal governments to orchestrate their withdrawal from their respective economic spheres.

The Logic of Integration

The shift away from the logic of counterweight to that of integration was gradual, and it was based on Ottawa's calculation of the changing external realities it was facing. More important still, the shift in favor of free trade amounted to nothing less than a watershed in the evolution of the commercial policy of Canada. In this section, I will focus on international factors, namely changes in Canada's vulnerability dependence and in its relative economic power position. The debate on free trade particularly focused on the systemic conditions facing Canada, gauging whether such conditions were working in favor of or against Canada. As such, the debate considered whether Canada could preserve its political independence and cultural identity, retain a reasonable degree of policy autonomy, and in general, whether it would be able to resist American domination under economic integration. In addition to these political concerns, the debate weighed the likelihood that a free trade arrangement would enable Canada to attain such economic goals as achieving industrial rationalization, enhancing national productivity and efficiency, and increasing economic growth, employment, and exports of manufactured goods.

The critics of free trade tended to view it as a costly arrangement, arguing that it would impose on Canada significant political and cultural costs that would outweigh any economic benefits derived from trade liberalization. One concern was that Canada would have to sacrifice its own interests in order to pursue continental ones, with no guarantee that those mutual goals would always produce equal benefits for Canada. As Barry (1984) notes: "Too much bilateral intimacy encourages [Canada] to subordinate legitimate industrial interests to common concerns" with the potential to inflame "anxiety about U.S. dominance." Other critics argued that a free trade arrangement would relegate Canadians to the role of "hewers of wood and drawers of water;" that the country would face rapid deindustrialization as American-owned branch-plants withdrew from Canada due to the diminished border-jumping incentive; and that the Canadian economy would become the dependent-periphery serving the American industrial heartland (Barlow 1990; Hurtig 1991).

Still, others viewed Canada's bid to create a free trade arrangement as an act of desperation, a policy initiative to which the country was drawn for no other reason than to minimize its losses in the future. As Williams (1987, 109) notes: "In the face of pressing 1980s' threats to Canada's economic stability posed by intermittent recession, high unemployment, international trade challenges from Japan, Europe and the newly industrialized countries, and the rise of protect-tionist forces in the United States, the free trade talks are essentially an attempt to adjust the balance point in the continental relationship so that the already existing network of commercial associations can be safeguarded."

In supporting free trade, Ottawa and the business community recognized Canada's improved international position and believed that economic integration would spur industrial adjustment and serve as a key pillar of Canada's future economic prosperity and political independence. According to Johnson, "growing wealth gives . . . an increasing capacity for individual self-fulfillment and the resources necessary for the achievement of national objectives" (quoted in Nossal 1985, 80). Similarly, the Royal Commission on the Economic Union and Development Prospects for Canada (Macdonald Commission) asserted, "In the modern era, one of the crucial components of satisfying Canadian identity is to be found in our economic capacity to perform at a high level in a competitive world"—which it was believed free trade would facilitate (1985, vol. 1, 62). Moreover, Ottawa and the business community believed that they could secure a bilateral arrangement that would create safeguards against potential trade disruptions stemming from heightened American protectionism.

Thus, from the point of view of the government and corporate Canada, the country had every reason to approach the issue of continental free trade with confidence. In a speech to the Strategic Planning Forum in October 1984, Secretary of State for External Affairs Joe Clark observed that "closer economic relations with the United States, if played right, can enhance [Canada's] voice and influence in international affairs. Canada possesses a new maturity as a nation [and] the modern purpose of Canadian nationalism is to express ourselves, not protect ourselves." While addressing members of the Economic Club of New York in December 1984, Prime Minister Mulroney observed that Canada's "status as a North American nation [was] a source of strength." Canadians were now "mature enough as a nation and confident enough . . . to recognize this reality and to take pride in an amicable relationship with a neighbour as powerful as the United States."

As the previous chapters have argued, the logic of counterweight was closely followed throughout the 1970s and early 1980s and reflected Canada's deteriorated position relative to the world and the United States, as well as a moderate level of vulnerability dependence. Defensive in nature and more tilted toward domestic efforts, this logic impelled Canada to intensify trade relations with Europe, Japan, and Third World countries and to fall back on renewed multilateralism in order to balance out its heavy dependence on the American market. Internally, this logic pushed the government to assert greater domestic control over the economy and to take a more active role in directing national economic development. The shift from the logic of counterweight to integration coincided with the rise of Canadian vulnerability dependence—brought about by an increase in relation-specific investments and trade dependence—and modest improvements in Canada's capabilities relative to the United States. These external factors combined to create part of the incentive structure that encouraged Ottawa to initiate free trade negotiations with the United States.

Although asymmetries in power capabilities between Canada and the United States have been large, the confidence the former demonstrated on the eve of announcing its decision to opt for free trade stemmed in part from the stability in some, and moderate improvements in other dimensions of Canada's relative economic capabilities. However, the very sources that underpinned Canada's improved position also worsened its vulnerability dependence relative to the United States. This has been the conundrum facing Canada since the 1980s. A number of important indicators demonstrate the positive changes in the standing of Canada in the 1980s and early 1990s. In general, Canada achieved modest yet significant gains in its share of the combined gross domestic products of the United States and Canada, from 10.71% in 1974 to 12.40% in 1985 (OECD 1987).

Another positive sign was the increase in Canadian control of domestic assets and the concurrent decline of American ownership. Canadian control of manufacturing activities increased from 39% in 1970 to 51% in 1984, while Canadian control in the oil and natural gas industries rose from 24% in 1970 to 56% in 1984 (Houle 1987, 486). This reflected the increase in Canada's ability to assert, through state and private efforts, greater control over national economic activities. This was particularly the case in the oil and natural gas sector where the Canadian share more than doubled from 1975 to 1984.

During the 1980s and the first half of the 1990s, the position of non-bank majority-owned U.S. affiliates in Canada grew somewhat weaker, while their branch-plant profiles vanished. Although their numbers increased, their share of Canada's GDP diminished from 11% in 1982 to 9.4 in 1989 to 8.5 in 1994, while their share of Canadian employment dropped from 7.5% in 1983 to 7.0 in 1989 to 6.2 in 1994 (Guillemette and Mintz 2004, 8). What is more, starting in the 1980s, the foreign direct investment in Canada (FDIC) proved to be more beneficial to the Canadian economy. Focusing on the manufacturing sector, Baldwin and Gu (2005) find that, in contrast to domestic-controlled firms, foreign-controlled plants have been more research and development-intensive, productive, innovative, technologically advanced, and inclined to pay higher wages and hire skilled workers.

Moreover, Canadian multinationals performed just as well as foreign-controlled firms in Canada. Of significance, foreign-controlled plants accounted for a disproportionate share of the annual labor productivity growth in Canadian manufacturing, which increased from 1.3% in the 1980-1990 period to 3.0% in the 1990-1999 period. The authors calculate that nearly 70% of the overall productivity growth for the two periods was owed to foreign-controlled firms, but the U.S. contribution decreased from 48% in the first period to 45% in the second. Moreover, of the 1.7 percentage point increase in productivity growth between the two periods, U.S.-controlled plants contributed 0.7 percentage points, other foreign-controlled firms contributed 0.4 percentage points, and Canadian-controlled firms contributed 0.6 percentage points (Baldwin and Gu 2005, 27). In sum, Canada reaped more advantages from the new generation of foreign capital, a development that played into Prime Minister Mulroney's liberalization of the foreign investment regime.

Canada's international investment position improved remarkably as the total amount of Canadian direct investment abroad (CDIA) more than tripled in size from $16.4 billion in 1978 to $54.1 billion in 1985 and nearly doubled from the mid-1980s to 1992 (Statistics Canada 1991, 42-3). The U.S. share of CDIA jumped from 53% in 1977 to 70% in 1985 and dropped to 58% in 1991. And, during this period, CDIA in the United States was more diversified than it was in other countries, with investments more or less evenly distributed among finance and insurance, communications, chemical, metal, and energy industries. As CDIA expanded, so too did the number of large Canadian multinationals. The number of Canadian firms owning more than $1 billion worth of foreign assets increased from ten in 1985 (accounting for 40% of total CDIA) to seventeen in 1991 (representing 50% of total CDIA) (Chow 1993, 11). Overall, corporate Canada became more internationally-oriented in the 1980s. And, while the initial wave of outward investment was met by weak earnings and financial losses, Chow (1993, 14) notes that "Canadian multinationals used that experience to learn the ropes of foreign markets, consolidating, restructuring and repositioning their activities and investments abroad."

The 1980s witnessed an important turning point in Canada's investment activities, as evidenced by the faster increase in the cumulative value of CDIA relative to FDIA. The persistence of the new investment pattern transformed Canada from a net capital importer to a net acquirer of foreign assets starting in the mid-1990s. According to Doran (1982-3, 133 and 138), the change in the investment pattern helped strengthen Canada's relative position to the extent that "Canadian firms had increasing control of assets abroad that could be regarded as a counterweight to foreign ownership and control in Canada." This and other changes in Canada's relative position, Doran contends, "increased Canadian influence relative to the United States, although marginally and slowly." Significantly, the more Canadian investors acquired assets in the United States the more concerned they became about protecting their investment against American protectionist threats and securing market access to the American market.

Although cross-border trade intensified between Canada and the United States in the 1980s and the first half of the 1990s, the asymmetry in the significance of two-way trade did not change. The U.S. share of Canadian merchandise exports rose from 62% in 1970 to 61% in 1980 to 73% in 1990 and to 80% in 1995. The Canadian share of U.S. merchandise exports dropped from 21% in 1970 to 16% in 1980, but rebounded to 21% in 1990 where it stayed relatively unchanged during the 1990s. The U.S. share of Canadian merchandise imports was 68% in 1980 (a one percentage point drop from 1970) and 67% in

1995 (a four percentage point rise from 1990). The Canadian share of U.S. merchandise imports moved from 28% in 1970 to 16% in 1980 to 18% in 1990 to 19% in 1995 (Molot 2003, 36). In general, whereas both countries saw their level of export dependence on one another increase in the 1980s and early 1990s, their level of import dependence remained largely unchanged.

As more industries on both sides of the border opted to split up stages of production and locate them in each other's market, the nature of two-way trade changed. Pioneered in the auto sector after the Auto Pact came into force, out-of-house cross-border fragmentation of the production process spread in the 1980s to other areas of the manufacturing sector (such as the agri-food, electrical/electronic, chemical, wood, metals and metallic minerals, and energy industries) as well as to the service sector (such as the finance and insurance industry). One of the indicators of cross-border integration of production is the incorporation of imports in making exports. Cameron and Cross (1999) show that the import content of Canadian exports in 1986 was 27.6% and rose to 32.4% in 1994. The export commodities with the highest import content were motor vehicles and parts, machinery and equipment, electronic equipment, plastics, textiles, and metals—whose primary destination was the United States. Considering that the United States supplied roughly 70% of the intermediate inputs integrated into these Canadian exports and that roughly 70% of these exports were destined for the U.S. market, two-way trade in the second half of the 1980s was becoming heavily intra-industry oriented and involved a significant amount of trade-specific investment by industries on both sides.

Another effect associated with the acceleration of cross-border production fragmentation was the rise in intra-firm trade. Cameron (1998, 18-23) estimates that U.S.-controlled companies operating in Canada received 61.5% of their total imports from U.S.-based affiliates in 1990, which increased to 66.5% in 1992. Of the total intra-firm exports to the United States between 1990 and 1992, U.S.-controlled Canadian affiliates accounted for 95.9% of the trade. Moreover, aggregate foreign-controlled intra-firm exports to the United States as a proportion of total foreign-controlled exports to the United States increased from 63.1% in 1990 to 69.7% in 1991. Intra-firm imports and exports were concentrated in high technology/high value-added Canadian manufacturing industries—namely, transportation equipment, electrical and electronic products, chemical and textiles, and machinery and equipment. Overall, Canadian-based subsidiaries of foreign multinationals accounted for about 51% of total Canadian imports and 44% of total exports in the early 1990s.

Canadian industries moved up the global value chain in the 1980s and made additional advances in the 1990s. This development was the result of the emerging pattern of cross-border investment and trade activities that contributed to the integration and specialization of production in key sectors of Canadian and American manufacturing and services. The economic advantages of having a larger share of Canada's industrial base linked to North American supply chains have been notable. Cameron and Cross (1999, 3) note that "the importance of trade to the [Canadian] economy does not come from an excess of exports over imports: rather, it is from the productivity gains that accrue with increased specialisation." Indeed, the productivity growth effect of specialize-tion was apparent in the second half of the 1980s—as exhibited by the increase in the value-added content of exports in GDP, which increased from 21.5% in 1986 to 26.3% in 1995. Moreover, the integration of key Canadian industries to North American supply chains, as Hart (2006, 126) observes, has meant that "as the U.S. economy moves further up the value chain, so does the Canadian economy."

Against the backdrop of the modest improvements in Canada's relative standing, was the inescapable reality that a growing share of the Canadian economy was deeply connected to the U.S. market. This development meant that the economic cost associated with disruptions to market access had risen to an unprecedented level for Canada. At issue was the increasing use of American trade remedy measures in response to industry complaints about the unfair trading practices of other countries.

Although such trade remedy measures had been used against Canada in the 1970s, it was the increase in the number of cases targeting Canada after the passage of the 1979 American trade legislation that raised fresh worries. The 1979 legislation exposed the bureaucratic trade remedy process to greater congressional pressures and also lowered the injury test that determined the eligibility of petitioners to receive countervailing duty support. According to Doern and Tomlin (1991, 68), American trade remedies were used fifty-one times against Canada between 1980 and 1986, marking a significant increase compared to the 1970s. Of these cases, more than 75% of the preliminary rulings by the U.S. International Trade Commission were against Canada. What was particularly troubling for Canada was the fact that the American trade remedy or contingency protection mechanism was becoming increasingly politicized and susceptible to the whims of protectionist forces brewing in the United States. "The problem was how [Americans] interpreted [their trade remedy laws], and what they did with it, and how responsive they were to the protectionists, and how arbitrary the handling of that legislation was," observed Simon Reisman, Canada's chief free trade negotiator (2000, 91).

The potential cost of using American trade remedies against Canada profoundly concerned Ottawa and Canadian exporters. As Mulroney's govern-ment pointed out in 1986:

> Last year protectionist threats and measures (quotas, anti-dumping and countervailing duties, surcharges) affect[ed] virtually all regions of Canada. Almost $6 billion of Canadian exports to the U.S. or 6% of the total were involved. The industries concerned provided some 146,000 jobs, a significant proportion of which were at risk (quoted in External Affairs Canada 1986, 1).

The insecurity that Canada faced as a result of American protectionist forces also had implications for Canada's emerging strategy of liberal continentalism.

The Conservative government was counting on increased access to the American market to induce domestic industrial rationalization, to achieve economies of scale that would improve Canada's international competitiveness in manufacturing, and to improve national productivity. In effect, Canada's efforts to redress its industrial adjustment problem through this strategy rested heavily on the security of Canadian access to the American market.

The above analysis of the two systemic forces sheds light on what external pressures and opportunities were at play as Ottawa changed industrial adjustment strategies in the first half of the 1980s. On the one hand, the strategy of continentalism reflected the newfound confidence among Canadians in their ability to engage directly with the United States. On the other hand, support for the strategy stemmed from the element of necessity, for the rise in American protectionism put at risk the Canadian-based relation-specific investments that had contributed to deepening cross-border production integration, which, in turn, was becoming a vital source of economic growth for Canada. Continental free trade, as Lipsey (2000, 102) observes, was therefore motivated by both defensive and offensive reasons. "The defensive reason was that the best defense against rising U.S. protectionism was to act bilaterally to eliminate all tariffs and to contain the use of non-tariff barriers." The offensive reasons for a free trade arrangement "were that the Canadian economy had matured and was ready to compete openly in world competition."

Domestic Determinants of Liberal Continentalism

Although systemic constraints and opportunities influence state preferences in a general direction, such an analysis only offers a partial explanation of actual state behavior. The implementation and consolidation of liberal continentalism was a product of the convergence of societal and governmental (both provincial and federal) preferences on market outcomes and the intervening effect of institutions. Although Conservatives were historically less supportive of trade liberalization than were Liberals, the Conservatives under the leadership of Joe Clark and Brian Mulroney were ideologically sympathetic to free market economics. As such, they advocated policies that promoted deregulation, privatization, and free market enterprise.

With their 1984 landmark electoral victory, the Conservatives used the apparatus of the state to achieve these goals. But support for free trade was not wholly secured until 1985. Launched by Prime Minister Trudeau in 1982 and chaired by Donald Macdonald, a former Liberal minister in Trudeau's Cabinet, the Royal Commission on the Economic Union and Development Prospects for Canada (Macdonald Commission) convinced Mulroney's Conservative government to abandon the party's historical position on trade liberalization and come down in favor of free trade with the United States. One of the key recommendations made by the bipartisan Macdonald Commission was the need to seek a free trade arrangement with the United States. In recalling the key factors that influenced Mulroney's decision to initiate free trade negotiations, former Chief of Staff to Prime Minister Mulroney Derek Burney (2000, 61) notes that "there can be no doubt . . . that the thorough analysis and virtually unanimous recommendations by Mr. Macdonald and his associates proved to be a major catalyst for the negotiation and a bedrock of faith for the political will in Canada."

Historically, the Canadian business community advocated free market enterprise, but until the early 1980s it opposed free trade with the United States. The business community's change of preferences was fueled by the 1981-82 recession, its apprehension about the intensification of economic nationalism in the Trudeau government, and its concern about American protectionism. Leading this historical change were the Business Council on National Issues (BCNI) and the Canadian Manufacturers' Association (CMA). In 1981, the BCNI created a trade policy committee to examine Canada's commercial strategy. By late 1982, the review led the BCNI to support the idea of negotiating a comprehensive trade agreement with the United States. For its part, the CMA changed its mind on the idea of a comprehensive trade enhancing agreement in 1983 after its members had been hit hard by the recession, saw an increase in their export activities, and recognized the need to improve their international competitiveness. By early 1985 the two business groups were joined by the Canadian Chamber of Commerce and the Canadian Federation of Independent Business in supporting a comprehensive bilateral trade agreement. Central to their position was the need to secure access to the American market and to roll back government involvement in the economy on both sides of the border (Doern and Tomlin 1991; Hart 1994; Leyton-Brown 1986).

Provincial governments constituted the third group of actors in the emerging coalition supporting the free market idea. By the mid-1980s most premiers, with the exception of Ontario's Liberal Premier David Peterson, supported a free trade arrangement with the United States. The four Western provinces—British Colombia, Alberta, Saskatchewan, and Manitoba—were early supporters of free trade. British Colombia Premier Bill Bennett and Alberta Premier Peter Lougheed were the strongest supporters, whereas Manitoba's New Democratic Party Premier Howard Pawley joined the other premiers after being reticent initially on the issue.

While western premiers favored free trade because it would secure access to the American market for their provinces' goods, their support also stemmed from another motive. Having been adversely affected by Trudeau's National Energy Program, Lougheed and his successor Donald Getty sought to use free trade to limit Ottawa's ability to intrude in the energy sector. The four Maritime provinces—Nova Scotia, Newfoundland, Prince Edward Island, and New Brunswick—were generally supportive of a free trade arrangement. During the negotiations, the four coordinated their efforts on issues that were particularly important to them, such as ensuring that the funds they received and depended on from regional development programs would not be subject to countervailing actions under the free trade accord.

Whereas Quebec moved cautiously and slowly in favor of free trade under the pro-sovereignty Parti Québécois, led by Premiers René Lévesque (1976-1984) and Pierre-Marc Johnson (1984-1985), Liberal Premier Robert Bourassa strongly supported free trade after winning the December 1985 election. Since 1985, both political parties have advocated free trade and, in general, greater international economic interdependence. Starting with Lévesque, Quebec governments became increasingly committed to the free trade idea as the Quebec business community grew more confident in its ability to compete internationally. Although there were a number of inefficient industries in Quebec, a growing number of firms, such as Bombardier, Cascades, Teleglobe, Vidéotron, and SNC, excelled in international markets. Security of access to the American market would allow other Quebec firms to achieve the economies of scale necessary to develop competitive advantages. In addition, added competition resulting from the reduction of tariffs would force Quebec's industries to rationalize and specialize, and thus improve their productivity and efficiency. Quebec's support of free trade reflected the province's increasingly export-oriented economy. In considering Quebec's decision to support a free trade arrangement with the United States, a Parti Québécois official noted:

> Protectionism is not an option when we export more than 50 percent of our GDP. We are a population of 7.5 million, our domestic market is small, and what we produce must be exported through vigorous efforts. Could Bombardier survive by simply relying on Quebec's internal market? It's absurd to even think about it.

In addition to this economic rationale, a political logic underpinned free trade. The two main political parties in Quebec have supported free trade as a way "to increase the relative autonomy of Quebec economic agents vis-à-vis the central government," as Meadwell and Martin (1996, 77) observe. However, the Parti Québécois has considered free trade to hold an even greater strategic value, for the economic integration of Quebec and United States "would help in easing the transition to independence and ensuring the economic viability of a sovereign Quebec."

Ontario was reluctant to join the other provinces in supporting a free trade arrangement. Ontario Premier David Peterson preferred the status quo, fearing that free trade would not produce enough gains to outweigh the high adjustment costs—plant closures and a rise in unemployment. Still, there were other reasons explaining Ontario's reservation. Stairs (1986, 60) notes that Ontario was holding back in order to gain more favorable concessions from federal negotiators. Also contributing to Ontario's intransigence was the fact that the provincial government believed that Mulroney advocated free trade just to consolidate his party's electoral base in Quebec and Western Canada, not because of its compelling economic logic.

From a domestic viewpoint, there was a political and economic rationale for pursuing the strategy of liberal continentalism. Mulroney committed his government to achieving national reconciliation after years of intergovernmental rivalry. In his November 1984 Speech from the Throne, the Prime Minister announced that his "government's management of federal-provincial relations will [aim] to harmonize policies of our two orders of government, to ensure respect for their jurisdictions, and to end unnecessary and costly duplication." Mulroney had placed much emphasis on this issue during the campaign leading to the September 1984 parliamentary election. "Our first task is to breathe a new spirit into federalism. The serious deterioration of federal-provincial relations is not exclusively the result of constitutional deficiencies. Centralistic and negative attitudes are much more to blame. Let us replace the bias of confrontation with the bias of agreement" (quoted in Simeon and Robinson 1985, 301-2).

The Tory government sought to avert the repeat of policy evolution via thrust and riposte dynamic between Ottawa and the provinces by rolling back state involvement in the economy. As the Macdonald Commission (1985, vol. 3, 147) observed, "Federalism seems to be the enemy of policy that is planned, comprehensive, coherent, uniform and consistent." It added, "In a country as diverse as Canada, centralization would be a recipe for paralysis," as was demonstrated during the strategy of economic nationalism. Market liberalism would improve intergovernmental relations and strengthen the internal economic union by removing those federal and provincial regulatory and commercial policies that were regionally or provincially biased or that intruded in the jurisdictions of other governments.

Moreover, it was believed that competitive federalism would help entrench market liberalism since governments would be inclined to emulate each other's market reforms if they produced desired outcomes. As the Macdonald Commission (1985, vol. 3, 148) pointed out, "In a world of uncertainty and rapidly shifting economic challenges where there is little understanding of what is likely to work best, [market-enforcing federalism] provides the opportunity for experiment and learning, for flexibility and inventiveness. It enhances sensibility to different viewpoints and permits canvassing of multiple sources of information and intelligence in different settings." Thus, intergovernmental policy coordination would be fostered via market policy experimentation and be largely voluntary.

The second objective of liberal continentalism was to facilitate the transformation of the manufacturing sector—which still had some branch-plant-like characteristics—into an internationally competitive one. As the Macdonald Commission (1985, vol. 1, 66) observed, years of state intervention had stifled industrial adjustment by creating "a number of market-distorting, growth-suppressing policies which redistribute[d] income to protected and privileged enclaves in the economy, reduce[d] economic efficiency, and inhibit[ed] flexibility." Central to the strategy of economic liberalism was the notion that privatization and deregulation, and other measures aimed at sharpening market processes, would encourage Canadian industries to rationalize and specialize, to increase research and development activities, and improve productivity.

Expanding Market Governance

The aim of the new adjustment strategy was to substitute market governance for state discretion. In setting the beat to which Mulroney's government would march, Finance Minister Michael Wilson pledged to shrink the size of the state after years in which the Liberals expanded the scope of state activities. "A major reason for our poor performance has been the failure of the Government of Canada to deal with the real problems. Through excessive regulation and intervention, it has substituted the judgments of politicians and regulators for the judgments of those in the marketplace" (Department of Finance 1984, 1-2). To bring about the withdrawal of the state from the economy, the federal and provincial governments employed such market instruments as deregulation, privatization, and free trade.

Deregulation: The Dismantling of the NEP

The objectives of the National Energy Program (NEP) in 1980 were to achieve security of oil supply, expansion of Canadian participation in the national oil and gas industry, and fairness in the division of energy benefits. A product of federal unilateralism, the NEP contributed to the expansion of federal involvement in the oil and gas industry by imposing strict regulations and a revenue-sharing scheme on energy producing provinces. Outraged by such a display of aggressive unilateralism, Alberta retaliated against Ottawa. Under significant domestic pressures, Ottawa relaxed some of the NEP's programs in late 1981 as it struck bilateral agreements with Alberta, British Columbia, and Saskatchewan. When Mulroney was the leader of the opposition in the Parliament in the early 1980s, he appointed Pat Carney to serve as the opposition energy critic and to develop the Conservative energy policy. The Conservatives called for terminating the NEP and devising an alternative plan more aligned with the interests of the provinces and the energy industry (Nemeth 2001). When the Tories assumed power, one of their first policy measures was to dismantle the NEP.

The dismantling of the NEP occurred in a piecemeal fashion through a series of federal-provincial agreements between 1985 and 1988. In 1985, three agreements were reached: in February, the Tory government signed the Atlantic Accord with Newfoundland; in March, the Western Accord was signed with Alberta, Saskatchewan, and Manitoba; and in October, the natural gas agreement was signed. In August 1986, the Canada-Nova Scotia Offshore Petroleum Resources Accord was approved, and in September 1988, the Northern Accord was signed with the two territorial governments—Northwest

Territories and Yukon. The Atlantic Accord gave Newfoundland the resource-ownership rights that Trudeau had forcefully withheld, which meant that the province would have the authority to manage offshore energy development and to tax offshore resources.

The Western Accord eliminated several federal taxes or special charges on petroleum that Carney described as "cruel and arbitrary discrimination" against western Canada and agreed to decontrol crude oil prices. The deal amounted to a loss of federal revenue, to greater relative autonomy for the provinces, and to an increase in industry profits. Similar to the Atlantic Accord, the agreements with Nova Scotia and the territorial governments devolved the functions of managing and taxing oil and gas resources. In sum, these individual energy-related arrangements between the federal government and provincial and territorial authorities terminated the NEP that Trudeau had unilaterally imposed on the provinces in 1980. Importantly, the decentralized approach to handling the energy issue implied that the Tory government "largely accepted a system of regionalized—meaning provincialist—systems of oil and gas management" and was aware that such a system would "became increasingly oriented along the North-South rather than the East-West axis" (Fossum 1997, 199).

Economic and political factors pushed the Tory government to dismantle the NEP. As far as economic factors were concerned, changes in the international oil market conditions over the past five years obviated the need for a national policy on energy. The rise in energy prices in the 1970s elevated significantly the strategic value of energy resources. The NEP and the energy-related mega-projects were designed specifically to harness Canada's potential energy capacity and make the country a major international energy producer. All of this changed, however, when energy prices began to drop in 1980. Not only did this change create a disincentive to engage in large capital investments to develop mega-projects, but it also lessened the need to ensure self-sufficiency and security of supply. Also, governments began to withdraw from their domestic energy sectors as the element of urgency fueled by the oil crisis subsided.

The Conservative energy policy also reflected Mulroney's desire to achieve national reconciliation. The bilateral arrangements were in part intended to defuse tensions between the two levels of government. They did so by curtailing the federal capacity to intervene in provincial jurisdiction, letting the energy sector be governed as much as possible by the operation of the marketplace, and respecting the interests of energy producing provinces. As Milne (1986, 99) observes:

> It was in the field of energy policy that the Mulroney government made its ideas about the mutually supporting goals of cooperative federalism and economic renewal most clear. It did so with dramatic new initiatives that reversed the direction of federal-provincial policies in this sector, diminished intergovernmental rivalry and competition, and cleared the way for private sector reinvestment.

If these bilateral arrangements initially ameliorated intergovernmental relations on the issue of energy by establishing new terms of engagement, the energy provisions in Chapter 9 of the 1988 Canada-United States Free Trade Agreement (FTA) helped entrench this pattern of relations. As Doern and Tomlin (1991, 80) point out: "For Alberta and other producer provinces in Western Canada, the FTA's energy provisions were seen as an insurance policy, not only against future U.S. protectionist actions, but even more against any more federal interference in provincial energy affairs. To them, the FTA offered an opportunity to forge an energy producers' charter of rights."

The FTA provisions prevented Ottawa and Washington from imposing price controls on exports and imports of energy products or discriminatory taxes. For example, neither country could export energy products at prices higher than what it sold them for in the domestic market. If an export tax was imposed on exports of energy products, then the tax also had to apply to domestic energy consumers. Moreover, a proportionality clause was included to ensure that in the event either country implemented regulatory measures, such as energy conserva-tion, which had the effect of reducing energy exports, domestic consumption would also be cut by a proportionate amount. In sum, the energy provisions of the FTA catered to the interests of both the oil and gas industry and the energy producing provinces by creating a set of rules that precluded the Canadian federal government from launching another national energy policy.

From 'Border Guard' to 'Genial Host': The Deregulation of Foreign Investment

In his 'Open for Business' speech before the Economic Club of New York in early December 1984, Prime Minister Mulroney discussed his government's plan to abolish the Foreign Investment Review Agency (FIRA) and establish Investment Canada—geared to attract foreign investment. "Our message is clear: Canada is open for business again," the Prime Minister asserted. "The government of Canada is there to assist—and not harass—the private sector in creating the new wealth and new jobs that Canada needs." Back home, the Investment Canada Act, tabled in the House of Commons by Industry Minister Sinclair Stevens on December 7, was supported by the provincial governments, the business community, and Canadians in general.

The creation of Investment Canada in late June 1985 put an end to the highly-controversial FIRA and contributed to a policy framework that promoted foreign investment and constrained state interference in this area. As Senior Vice-President of the Canadian Chamber of Commerce Roger Stanion observed, "The government has replaced FIRA, the border guard, with Investment Canada, the genial host to responsible visitors" (quoted in Globe and Mail 1984, B6). Investment Canada marked a significant departure from the FIRA. The new agency was based on the idea that foreign-owned affiliates could be trusted to help strengthen Canada's technological capabilities, create new employment opportunities, and improve national productivity without compelling them via regulatory measures.

Initially, Investment Canada reported to the Department of Regional Industrial Expansion (DRIE), whose minister was responsible for making a final decision with respect to review approval. However, in 1987 the agency was subsumed under the Department of Industry, Science and Technology Canada (ISTC), which was formed out of a merger between the DRIE and the Ministry of State for Science and Technology (MOSST). Mulroney considered the ISTC to be the government's "flagship" department to oversee microeconomic policy, and envisioned Investment Canada as playing a facilitating role in this policy arena. As such, the new department unveiled three "cornerstone programs": (1) the strategic technologies program, which sought to foster information technology, biotechnology, and advanced materials production such as metals, alloys, ceramics, polymer, and composites; (2) the sector competitiveness initiatives, which were designed with a view to enhance international competitiveness; and (3) the provision of business information and development services. Along with other departmental instruments, the organizational resources of Investment Canada were to be deployed to help advance these goals (Doern 1990).

The foreign investment review mechanism of Investment Canada differed from that of the FIRA. Investment proposals were evaluated on the basis of "net benefits," rather than the FIRA's elusive "significant benefits." Under this new standard, incoming foreign investments were approved so long as they brought some kind of benefit regardless of the significance of such a benefit. As Jenkins (1992, 119) notes: "If there is one iota of benefit in an investment, it [had to] be approved." Moreover, Investment Canada's scope of review was much narrower. Unlike the FIRA, the establishment of new businesses by non-Canadians was not subject to review. However, reviews were required in cases when non-Canadians acquired direct control of Canadian businesses with assets worth $5 million or more. Reviews also applied to cases when non-Canadians acquired indirect control of Canadian businesses with assets worth $50 million or more as a result of merger arrangements or the acquisition abroad of the parent company of a Canadian subsidiary. In short, these and other provisions made Investment Canada a promoter of foreign investment. Moreover, the agency and ISTC were central in promoting the Tory government's firm-specific industrial policy, which stressed market-based incentives to improve productivity and competitiveness.

A number of factors contributed to the abolishment of the FIRA. First, societal preferences shifted more decisively in favor of foreign investment than ever before. A mid-1980s survey on Canadian attitudes toward U.S. investment in Canada indicated that, in contrast to the 1970s, fewer Canadians were of the opinion that there was too much U.S. investment (Bashevkin 1991, 96). Moreover, as LeDuc and Murray (1989, 134) point out, this change in Canadians' perception of American investment tended to be manifested relatively equally across Canada, with Quebec registering the greatest support for foreign investment and the Atlantic region expressing the least support. The Canadian business community was firmly in support of an open foreign investment regime. The 1979 Final Report of the Export Promotion Review Committee (Hatch Report) recommended that Canada practice "a more liberal policy towards direct investment in Canada." In underscoring one of the benefits of foreign investment, the report (1979, 17) noted that "some subsidiaries [had] developed unique products in Canada through which they [had] earned worldwide or regional marketing rights within their firms."

Indeed, if there was a common theme expressed by the business community with respect to foreign investment it was that Canada was beginning to exhibit fewer structural deficiencies attributed to a branch-plant economy, which had served as the justification for imposing foreign investment regulations in the first place. In the hearings on the Investment Canada Act, Robert Blair (1985, 8:5) of NOVA Corporation stated that "[i]n the NOVA company group, which is Canadian owned, we exist daily in competition around the world with some of the toughest international business rivals in several of the world's most competitive sectors." Thomas D'Aquino (1985, 11:28), President of the Business Council on National Issues, remarked that the realities of the early 1980s hardly resembled the "long history of branch plant development in this country." The prevailing view within the business community was that foreign investment neither hindered industrial development nor reduced economic growth, but rather had the opposite effect on both. In addition, more foreign-owned Canadian subsidiaries were exporting, marking an important departure from their once dominant import-oriented and domestic market-oriented practices.

Another factor contributing to the creation of Investment Canada was the need to improve the investment environment domestically. In its annual meeting in 1984, the World Economic Forum ranked Canada the least hospitable market for foreign investors out of twenty-two advanced industrialized economies. Although the National Energy Program had contributed to such a poor ranking, it was believed that the FIRA was equally to blame. Investment Canada helped improve Canada's image by significantly narrowing the scope of reviews of foreign investment proposals and substituting the judgment of the private sector for that of the state with respect to investment decisions. Furthermore, Investment Canada became an instrument with which the government could actively participate in developing investment opportunities for both foreign and domestic investors.

Finally, Investment Canada was designed to encourage more Canadian subsidiaries of foreign companies to develop world product mandates. As the Hatch Report pointed out, more foreign-owned subsidiaries operating in Canada in the late 1970s were implementing a global product mandating strategy in which their parent companies granted them nearly exclusive responsibility for the development, manufacturing, and marketing of particular products on a worldwide basis.

The implementation of a product mandating strategy can confer upon host economies vast benefits. Such benefits include increasing research and development activities; enhancing productivity and efficiency given longer production runs stemming from access to international markets that parent companies facilitate; contracting of local suppliers; and improving the managerial and technical caliber of the domestic labor market (Atkinson 1985). In view of these benefits, the Ontario Advisory Committee on Global Product Mandating (1980, 4), consisting of senior executives from Canadian multinational affiliates, noted that "governments can influence more multinationals to evolve to specialized missions in their Canadian operations. The most effective way to achieve these objectives is to emphasize performance rather than ownership of the multinational affiliate." In particular, the Advisory Committee asserted that the most effective way to attract this type of foreign investment was to improve the investment climate by minimizing government obstruction.

Like the deregulation of the energy sector, the deregulation of the foreign investment regime in Canada was further institutionalized with the inclusion of an investment chapter in the Canada-United States Free Trade Agreement. Canada was reluctant to discuss the investment issue in the bilateral trade negotiations simply because it did not recognize additional benefits in deregulating beyond the scope designated by Investment Canada; however, by 1987 Ottawa and Washington agreed to extend the principle of national treatment to each other's investors seeking to establish new businesses in either country. Moreover, Chapter 16 of the FTA established that Canadian provinces and American states were subject to provisions on investment. Under the FTA, Investment Canada was to increase gradually the review threshold for direct acquisitions from $25 million in 1989 to $150 million by 1993 as well as for indirect acquisitions from $100 million in 1989 to $500 million in 1991 and to end such reviews by 1992. Canada, however, did retain the right to review American investments in Canada's cultural industries and to establish restrictions with respect to the sale of Crown corporations.

The deregulation efforts of the Mulroney government in the energy sector and investment area were met with success chiefly because the existing institutional structure proved amenable to such adjustment policies. For one thing, deregulation reinforced the prevailing arm's-length pattern of relationship between business and the state by reducing the scope of state control and enhancing the autonomy of the private sector. The prevailing pattern of intergovernmental relations was amenable to deregulation, facilitating the withdrawal of the federal state from a policy area in which its involvement had been challenged by the provinces. Lastly, the process of deregulation eased the administrative and managerial burden on a bureaucracy that proved ineffective in attaining coordination from within.

Privatization: Dismantling or Rationalizing the Public Sector?

The Tory government's privatization achievements by the 1990s were mixed. Although privatization was an important adjustment policy of Mulroney, as suggested by the various policy pronouncements, the outcome fell short of what was anticipated. After becoming leader of the Progressive Conservative Party in 1983 (the official opposition party at the time), Mulroney established a task force to look into how the federal government could divest itself of Crown corporations. Like the task force headed by Pat Carney that was set up to study the deregulation of the energy sector, the privatization task force was given the mandate to formulate policy guidelines that the Tories would follow in the event they won the parliamentary election in September 1984. Although the Tories were exploring the privatization issue, the issue was not a focus of debate during electoral campaign. In early November, Mulroney disclosed the task force report to his cabinet, a few days after Industry Minister Sinclair Stevens announced that Ottawa was planning to sell five Crown corporations (Stanbury 1988).

Between late 1984 and May 1987, as Ottawa sold Crown corporations, the Conservative government made a number of policy statements that helped define the rationale for proceeding with privatization. Because public opinion had turned against the Clark government when it attempted to sell Petro-Canada, de Havilland, Canadair, and Eldorado, the Conservative government worked hard to mobilize public support. Initially, the government defended privatization using standard neoclassical economic arguments. Stevens remarked that privatization was a way for the "discipline and vitality of the marketplace [to] replace the often suffocating effect of government ownership" (quoted in Laux and Molot 1988, 191). Echoing Steven's message, Robert de Cotret (1985), President of the Treasury Board, observed that privatization would "reduce the size of government in the economy and make room for private sector initiatives; improve firm efficiency through market discipline by reducing political and bureaucratic impediments; and encourage investment by Canadians through the direct participation in the ownership of major national corporations which they have supported as taxpayers."

In an attempt to enhance the management of the privatization portfolio and the communication of the government's privatization intentions, Mulroney created in May 1985 a task force that included the cabinet ministers from the Treasury Board, Industry, Energy, and Finance. The task force was given the mandate to report directly to the Prime Minister on proposed privatization plans. In addition, in mid-1986, Mulroney reorganized the bureaucratic apparatus overseeing privatization by naming Barbara McDougall the first Minister of State for Privatization and Regulatory Affairs and adding the Privatization and Regulatory Affairs committee to the cabinet committee structure.

In 1987, the Office of the Minister of State for Privatization explained the government's motive for pursuing privatization. First, the government argued that it was necessary to privatize those Crown corporations whose "public policy goals [were] no longer legitimate" and whose "original objectives [were] no longer valid." Second, the private sector, it was argued, provided services more efficiently to customers than public sector actors. Third, in the face of fiscal constraints, the federal government was unable to continue to operate existing Crown corporations or channel additional federal funds into existing Crown corporations to help them survive. Fourth, subsidized Crown corporations created unfair competitive situations for the private sector, which contradicted the government's efforts to expand market governance. Finally, the government argued that Crown corporations were slow to adapt to external changes and were risk-averse, two behavioral conditions that undermined their ability effectively to compete and succeed.

In effect, the Mulroney government was seeking to replace the public enterprise culture that historically had legitimized the employment of state-owned enterprises to accomplish political-economic goals. As McDougall noted, "We are mindful of the fact that many of our national Crown corporations have earned the respect of Canadians for the key roles they played in the industrial and regional development of our great country." However, "[t]oday, Canada has a developed and maturing social and economic infrastructure;" rather than relying on the galvanizing force of the state, "the private sector is [now] recognized as being more appropriately the principal stimulant of economic growth and renewal" (Office of the Minister of State for Privatization 1987). Indeed, like deregulation, privatization reflected Ottawa's determination to substitute market governance for state discretion and to institutionalize market processes so that they become the principal basis of coordination.

In 1984 the federal government owned sixty-seven Crown corporations that in turn owned 128 subsidiaries. The bulk of these enterprises operated as commercial corporations, engaged in the production and sale of goods and services. The *Financial Post 500* indicated that nineteen federal Crown corporations were among Canada's 500 largest non-financial enterprises in 1985. According to the Economic Council, the provinces owned 203 parent Crown corporations that in turn owned 187 subsidiaries. In 1985, the *Financial Post 500* listed sixteen provincial Crown corporations among the top 500 industrial companies in Canada. Between 1984 and 1988, the Mulroney government undertook thirteen privatizations and dissolved another eight companies. Among the privatizations completed in the 1990s by the Tories and Liberals, five stand out for having generated large sale proceeds. Particularly noteworthy is that the Liberal government of Prime Minister Jean Chrétien oversaw the largest privatization plan ever when it sold the Canadian National Railway in 1995.

The Tory government's privatization strategy, however, did not always conform to its policy declarations. In fact, privatization did not necessarily contribute to greater market competition, withdrawal of the state, and strengthening of the private sector. In two particular cases the federal government engaged in lateral transfers by selling its corporate holdings to provincial governments. Ottawa sold the assets of the Northern Canada Power Commission to the territorial governments of Yukon and of the Northwest Territories in 1987 and 1988, respectively.

Ottawa again used lateral transfer in 1988 when it sought to offload Eldorado Nuclear Limited, the largest uranium producer in North America. Lacking a private buyer, a deal was negotiated with the Saskatchewan government in which the provincial Crown corporation, the Saskatchewan Mining Development Corporation, would buy the federal Crown corporation. A new company, Cameco, was created from the merger of the Saskatchewan Mining Development Corporation and Eldorado Nuclear Limited. It was agreed that Cameco would be gradually privatized; however, after a share offering to the public in 1993, the government found itself still owning 42% of the provincial Crown corporation (Laux 1993).

In other cases, the divestment of federal state assets resulted in the concentration of market players in particular industries. This was the case when the Mulroney government sold de Havilland Inc. to the American giant, the Boeing Company, which resulted in increased concentration in the Canadian and international aircraft manufacturing markets. The Quebec-based transportation equipment manufacturer, Bombardier, also augmented its market power in a series of purchases of Crown corporations. In 1986, Bombardier purchased Canadair, and when Boeing opted to sell the former Crown corporation de Havilland in 1991, the Quebec-based giant became the new majority owner of the company (Globe and Mail 1992). The company grew even larger when it bought from the Ontario government UTDC, a mass transit company (Globe and Mail 1991).

Bell Canada Enterprises—which ranked second in the *Financial Post 500* in 1988—benefited from the federal government's sale of Teleglobe Canada in 1987. Although Teleglobe Canada was sold to the Montreal-based Memotec Data Inc., Bell Canada Enterprises gained indirectly because it owned 27% of Memotec Data Inc. This particular trend in Canada's privatization process has been attributed to the fact that the federal government as been willing to sell its corporate holdings to single buyers (Schultz 1988; Stanbury 1988).

As the evidence suggests, the outcome of a privatization plan did not always amount to the complete elimination of state influence over the operations of the private sector. In many cases, foreign ownership restrictions were attached to privatization plans, especially after Ottawa faced criticism for selling de Havilland to Boeing. For example, the privatization of Teleglobe Canada included terms that prohibited foreign telecommunications companies from purchasing shares in Teleglobe and prevented non-residents from collectively owning more than 20% of the telecommunications company. Similarly, when Air Canada was privatized in 1988, a provision was added that prevented non-residents from collectively owning more than 25% of the company. As the deal securing the merger of Eldorado Nuclear Ltd. and the Saskatchewan Mining

Development Corporation was being completed, it was agreed that the aggregate share of Cameco held by non-residents could not exceed 20%.

In a few cases, the government displayed favoritism toward companies that had purchased Crown corporations. For example, after Teleglobe was privatized, its new private sector owner secured a deal that enabled it to maintain Teleglobe's monopoly position on the provision of satellite-based telecommunications services for five years. Shortly after Bombardier acquired Canadair, Ottawa awarded the Montreal-based company a lucrative contract to overhaul its CF-18 fighter aircraft despite receiving a lower bid by the more experienced Bristol Aerospace, based in Manitoba (Laux 1993; and Stanbury 1988).

Privatization also took place at the provincial level since the efficiency, fiscal, and political motives that underpinned the privatization of federal Crown corporations also applied at this level. However, provincial governments have also sought to hold on to Crown corporations and forcing them to be more "businesslike." Molot (1988, 415 and 417) points out that the provincial governments have faced obstacles that have delayed, slowed, and limited privatization:

> These [obstacles] include the economic realities of resource-based provincial economies, the relatively fewer number of economic levers available to provincial governments than to Ottawa which has resulted in their greater reliance on more direct instruments of intervention, continuing public support for Crown corporations, and the problems of finding interested (and appropriate) buyers.

Nevertheless, the provinces have proceeded with privatization. Between 1983 and 1988, seven provinces (British Columbia, Alberta, Saskatchewan, Manitoba, Ontario, Quebec, and Newfoundland) privatized twenty-four Crown corporations. Quebec accounted for one-third (eight) of the twenty-four privatizations while British Columbia accounted for seven. Although all ten provinces engaged in privatization in the 1990s, the pace of privatization varied. Alberta undertook several large scale privatization plans in the 1990s, including the sale of the Alberta Government Telephones, the Alberta Energy Company, Syncrude Canada, Vencap Equities Alberta, and Edmonton Telephone. Ontario's privatization program moved slowly and sought opportunities to enter into public-private partnerships as ways to downsize the province's public sector. Moreover, in contrast to Quebec's sweeping privatization efforts in 1986 and 1987, subsequent Liberal and Parti Québécois governments opted to proceed more slowly with privatization.

Like deregulation, privatization was a policy action that demonstrated the Canadian state's desire to strengthen the market while limiting the scope of state involvement in the economy. Unlike deregulation, however, privatization was not directly influenced by continental integration, which is why Ottawa and the provincial governments pursued privatization at different speeds.

Continental Integration: Negotiations and the Role of Provincial-Federal Consultations

The policy shift that led Canada to enter into a free trade arrangement with the United States began in 1983 with the release of a report by a task force in the Department of External Affairs (DEA). The consultations that the task force conducted with representatives of the business community and provincial governments in 1982 revealed that there was a growing concern about the state of Canadian-American trade relations. Canada's sense of vulnerability to disruptions of access to the American market grew more intense as the country's external trade became more bilateral and the economy became more dependent on trade.

Although the report emphasized the importance of improving the international trade regime under the General Agreement on Tariffs and Trade, it also raised the option of pursuing sectoral free trade with the United States, as Canada had done in previous decades with regard to the automobile industry, defense materials, and agricultural machinery. In June 1984, Trudeau opted to proceed with the bilateral sectoral free trade option and appointed Tony Halliday to negotiate sectoral trade agreements with his American counterpart, William Brooks, the United States Trade Representative. Given the lack of interest in the United States for such arrangements, the bilateral talks went nowhere.

In February 1985, International Trade Minister James Kelleher released a discussion paper, *How to Secure and Enhance Canadian Access to Export Markets*, which argued that it was imperative for Canada to find a way to enhance and secure commercial access to the United States. Several options were entertained in the report: (1) to support the status quo; (2) to negotiate sectoral or functional arrangements (for example, in government procurement and in contingency protection); (3) to negotiate a comprehensive bilateral trade arrangement; or, (4) based on the recommendation of the Business Council on National Issues, to negotiate a framework agreement that would improve and expand trade relations and establish an institutional mechanism to address trade issues and resolve disputes (DEA 1985, 24-8).

The drafting of the report did not go smoothly. Whereas Kelleher and External Affairs Assistant Deputy Minister for the United States Derek Burney favored the free trade option, others within the DEA opposed this initiative fearing that free trade would undermine Canada's political autonomy and its reputation as a supporter of the GATT. Opposition also came from officials in the Department of Regional Industrial Expansion who were concerned about the ability of Canadian industries to compete against their American counterparts under a free trade arrangement. In the end, Burney, who was in charge of drafting the report, muscled through the barrage of bureaucratic opposition and made the option of a comprehensive free trade a key component of the report. The report dismissed the first two options, for neither would take the bilateral relation in a direction that would serve the interests of Canada, particularly in securing and enhancing market access. Between the last two options, the report sided with the option of negotiating a comprehensive bilateral trade arrangement, arguing that the fourth option would be "essentially declaratory and non-contractual in nature" (DEA 1985, 28).

The momentum to achieve a bilateral free trade arrangement with the United States intensified when Mulroney and U.S. President Ronald Reagan held a summit meeting in Quebec City in March 1985. In a joint declaration, the two leaders made known their commitment to establishing "a climate of greater predictability and confidence for Canadians and Americans alike to plan, invest, grow and compete more effectively with one another." In working toward this goal, the two leaders agreed to find ways to lower barriers to trade and resolve trade-related disputes. To address the former issue, bilateral talks were initiated between Kelleher and U.S. Trade Representative Brock to outline ways to lower and eliminate existing trade barriers and to report with recommendations in six months. On the latter issue, the two leaders set in motion a process aimed at resolving disputes in eight specific areas. Of great importance for the Mulroney government was the Quebec City Summit revelation that Washington was interested in negotiating a deal with Canada to deepen trade relations.

Released in early September 1985, the Macdonald Report helped convert Mulroney to an advocate of free trade. The report articulated a convincing argument that held that free trade would not only facilitate industrial adjustment but would also offer a way to ameliorate intergovernmental relations. Moreover, the Commission's Report was released at a time when provincial governments and the Canadian business community favored a free trade arrangement, which only emboldened Mulroney to side with free trade. This was the message communicated to the Prime Minister by Kelleher who, since February 1985, had been conducting cross-country consultations on the issue with representatives of the Canadian business community and labor. This was also the message heard from provincial leaders during the Regina First Ministers' conference in February.

A few days after the release of the Macdonald Report, the Prime Minister announced in the House of Commons that the government would pursue the comprehensive free trade option. In late September, the Prime Minister informed the House of Commons that he had called President Reagan to commence trade negotiations. In a formal letter sent to Reagan on October 1, the Prime Minister called for the "two governments [to] pursue a new trade agreement involving the broadest possible package of mutually beneficial reductions in barriers to trade in goods and services," and that "such an agreement secure and enhance access to each other's markets . . . and result in a better and more predictable set of rules whereby [bilateral] trade is conducted" (quoted in DEA 1986, 77).

Canadian negotiators entered negotiations with three objectives. The first objective was to shield Canadian exports to the U.S. market from American protectionism. This goal was underscored on several occasions by Prime

Minister Mulroney and the negotiating team. "Our highest priority is to have an agreement that ends the threat to Canadian industry from U.S. protectionists who harass and restrict our exports through the misuse of trade remedy laws. Let me leave no doubt that first, a new regime on trade remedy laws must be part of the agreement" (quoted in Leyton-Brown 1988, 168). Second, Ottawa sought to enhance Canada's market access by deepening and broadening trade liberalization. Finally, Canadian trade negotiators sought to establish specific rules and procedures designed to adjudicate in a predictable, consistent, and fair manner disputes relating to trade practices. The goal was to develop "a strong dispute settlement mechanism to reduce the disparities in size and power and to provide fair, expeditious and conclusive solutions to differences of view and practice" (DEA 1986, 3-4).

The Canada-United States Free Trade Agreement (FTA), which came into effect on January 1, 1989, was a worthwhile deal for Canada. It satisfied most of the objectives Ottawa was seeking to attain. Canada's relative success in the negotiations was largely owed to the negotiating team's ability to command enough political and bureaucratic resources to make up for the country's middle power disadvantage. Also, Canada exhibited a higher level of focused attention in the negotiations, which in turn allowed for quick decision-making and systematic pursuit of specific goals. This was attributed to the fact that free trade proved to be a much more significant policy matter for Canada than it was for Washington. As Clayton Yeutter (2000, 75-6), U.S. Trade Representative under Reagan from 1985 to 1988, notes: "Although this was a big deal in both countries, it was a really big deal in Canada . . . , whereas in the U.S. we had a lot of other [foreign policy matters, both political and economic in nature] on our table at the time."

The trade liberalization component of the bilateral agreement helped enhance Canadian access to the American market, as illustrated by the increase in the volume of trade in liberalized sectors. Enhanced access created additional positive effects. For example, the free trade arrangement induced greater specialization and rationalization in continental business activities, a process that led to increased intra-industry trade and intra-firm trade in the 1990s. Moreover, enhanced access helped fuel much of the growth during the decade.

The free trade negotiations did address the issues of American trade remedy and dispute settlement. Although Canada had to settle for a compromise deal on trade remedy, it was one that still fulfilled the Canadian objective of securing market access. Canada's chief objective in the negotiations was to limit the ability of the U.S. government to employ its contingency protection laws, namely, countervailing duties and antidumping. In fact, the Canadian negotiation team made it clear that even if the negotiations resulted in enhanced access, a free trade arrangement would not be worthwhile if it did not include an institutional mechanism that allowed Canada to challenge American trade remedy practices. As the Canadian chief negotiator Simon Reisman (2000, 91) points out: "What's the point of entering into an agreement when as soon as you are successful, some protectionists—and in the U.S. you'll get lots—are going to be able to prevail on the Administration . . . and get things done that interfere with our trade."

To achieve this, the Canadian negotiating team sought to establish bilateral rules that clearly specified what constituted unfair trade practices and apply such standards to bilateral trade. Moreover, in the event that problems of interpretation of these rules arose, redress would be achieved through impartial, binding arbitration. In the end, the United States proved unwilling to substitute a common set of rules on fair and unfair trade practices for its own trade remedy laws.

The compromise deal led to the development of a dispute settlement institutional framework embodied in Chapters 18 and 19 of the FTA. Chapter 19 created a unique institutional mechanism for resolving disputes over countervailing duties and antidumping. The two countries agreed to employ their respective trade remedy laws, but also agreed to replace domestic judicial review of countervailing duties and antidumping decisions by a binational review panel. Composed of two panelists from each country and one panelist chosen jointly, the binational review panel was given the mandate to review and reverse a domestic decision if it was determined that a particular trade remedy action was based on an improper application of existing trade laws in the country where the case originated. Chapter 19 also included a provision requiring that Canada be specifically named in any changes made to the American antidumping and countervailing duty laws if they were to apply to Canada. Finally, changes to existing American trade remedy laws that named Canada were subject to the review of a binational panel to determine whether the legislative changes conformed to American obligations in the General Agreement on Tariffs and Trade and to the rules of the bilateral free trade agreement.

Chapter 18 of the FTA established a mechanism to resolve disputes concerning rights and obligations of the two countries under the FTA, with the exception of disputes linked to antidumping, countervailing duties, and financial services. If a dispute was not resolved through information sharing and direct consultations, the matter could be referred to the Canada-United States Trade Commission. In the event the Commission was unsuccessful in resolving a dispute, the case, among three possible procedures, could be referred to an ad hoc panel whose recommendations would be implemented by the Commission. The institutionalized arbitrative mechanisms that Chapters 18 and 19 embodied amounted to a significant change in how the two countries would manage their bilateral trade relations in the future. As Stanley and Blank (1999) point out: "The most substantial gains from the FTA [for Canada] may well turn out to have been in trade jurisprudence. Getting a trade partner as important as the United States to accept the principle that its own trade-remedy judgments should be 'internationalized' in a binding mechanism made up of experts from each country was extremely significant."

Provincial-Federal Consultations during the Free Trade Negotiations

The Canadian federal structure played an important intervening role between Ottawa's pursuit of its instrumental goals in the free trade negotiations and the substantive outcome of the negotiations. The authority to negotiate and sign an international trading arrangement has always rested with Ottawa. However, the national government's freedom of action in this jurisdictional area is limited because this authority does not translate into an ability to implement those provisions of international agreements that fall within provincial jurisdiction. As such, if Ottawa sought a free trade agreement that dealt with issues such as investment, financial and professional services, government procurement, deregulation of liquor monopolies, or subsidies, Ottawa would have to collaborate with the provinces in order to secure their support and compliance. In recognizing the jurisdictional limitations facing Ottawa, the Macdonald Report recommended "close consultations about provincial, as well as federal, objectives" before negotiations were launched, and "binding commitments to achieve them" during the negotiations, which was to be promoted by "sustained and continuous consultations by federal and provincial Ministers."

The issue of provincial involvement was addressed during the Annual First Ministers' Conference in Halifax in late November 1985. Alberta's Federal and Intergovernmental Affairs Minister Wayne Clifford (1991, 130) summed up the position of the provinces on the need to be engaged with Ottawa. "There has been an increasing realization by all provinces of the impact of changes in the world economy on provincial economies and the realization that provincial areas of jurisdiction fall increasingly within the scope of trade negotiations. What all of this has brought us to realize is that we have to work together, and that we need . . . consultation procedures." For Ottawa, provincial-federal consultations served several purposes. Provincial support was necessary in order to implement effectively any free trade deal. Also, Ottawa's interest in these consultations stemmed from Mulroney's desire to renew federalism—that is, reverse the tide of Western alienation and avert provoking sovereigntists in Quebec. The consultations would give the provinces voice opportunities through which they would be able to influence Ottawa's course of action. Finally, by securing the support of the provinces through consultations, Ottawa would be able to count on them to help sell the free trade agreement to the Canadian people.

At the conclusion of their November meeting in Halifax, the Premiers and the Prime Minister issued a general statement affirming their commitment to the "principle of full provincial participation in the forthcoming trade negotiations" (quoted in Barrows and Boudreau 1987, 140). However, for the next six months, wrangling rather than consensus emerged between Ottawa and the provinces. As expressed in a letter from Donald Getty, Premier of Alberta sent to Mulroney in

March 1986, the provinces had come to interpret "full provincial participation" to mean that provincial representatives would be directly involved in the negotiations, that the chief negotiator would receive his orders from provincial and federal governments, and that the provinces would have a veto on issues pertaining to their jurisdiction. As Ottawa was preparing for the negotiations, the provinces felt they were being denied a role commensurate with their authority.

Appointed to head the Canadian negotiating team in early November, Simon Reisman made it clear that the provinces would not be present at the negotiating table and that Ottawa alone would issue instructions for the negotiating team. However, Reisman was not against the concept of consulting with the provinces on issues that fell within their jurisdiction. He recognized that to ensure full implementation and compliance of a free trade agreement, it was necessary to gain the consent of the provinces prior to signing an agreement. With this in mind, Reisman began chairing monthly meetings in January 1986 with provincial advisors, which became known as the Continuing Committee on Trade Negotiations (CCTN). As a prenegotiation forum, the meetings were used to exchange views and gain a better understanding of each other's expectations and concerns. But this initiative did nothing to solve the impasse concerning the role that the provinces would play during the negotiations.

A resolution was reached during the First Minister's meeting in Ottawa in June 1986. It was agreed that Chief Negotiator Reisman would be accountable to the federal cabinet, that the Trade Negotiations Office (TNO) would be supervised by Reisman, and that no provincial representatives would be in the TNO or at the negotiating table. Second, in consultation with the provinces, Ottawa would formulate the mandate of the Chief Negotiator. Finally, in addition to having a First Minister's meeting every three months for the duration of the negotiation process for the purpose of reviewing progress in the negotiations, provincial trade representatives and Reisman would meet regularly through the CCTN to exchange information and views. Thus, the provinces would be fully consulted and briefed regularly, but they would not share negotiating authority with the federal government (Brown 1988).

From the start, the Western provinces took full advantage of this intergovernmental apparatus, as did Quebec. Ontario's involvement was less than enthusiastic. Unable to prevent free trade negotiations from going ahead, Ontario acted more as a spoiler than a constructive shaper in the series of intergovernmental consultations. Interestingly, as Ontario became more removed, Quebec's role and influence in the consultations increased, much to the satisfaction of the Bourassa government. The general feeling among most provinces after a free trade deal emerged was that the consultations that ran alongside the actual negotiations proved to be a major milestone in the way in which the two levels of government worked with each other in international negotiations.

Aside from the interruption of the regular provincial-federal consultations at the end of the negotiations, the consultations were successful to the extent of keeping the provinces informed and involved. From September 24 to October 4, 1987, formal intergovernmental consultations did not take place as the bilateral trade negotiations entered a period of brinkmanship in which Canada threatened to walk out of the negotiations if no agreement prevailed on the issue of trade remedy. When the *Elements of the Agreement* were released on October 4, outlining the agreement's basic provisions, the provinces did not find new issues that involved their jurisdiction. Overall, the free trade deal intruded less into provincial jurisdiction than was first anticipated by the provinces. As it turned out, provincial procurement, provincial technical standards, and financial services were not affected and only future provincial investment laws would be (Brown 1988).

When the final legal text of the free trade agreement came out in December, Ontario, Manitoba, and Prince Edward Island opposed the accord. For the Mulroney government, what was most important in order to ensure the approval of the agreement was that it pass the general constitutional amendment formula requiring parliamentary approval and the support of at least seven provinces with a combined population of more than 50% of Canada's. With respect to the latter, Mulroney could rely on the seven provinces that approved the free trade deal, among which Alberta and Quebec were its most enthusiastic supporters. As for the former criterion, he faced a more challenging task. When the Tory government tabled Bill C-130 (An Act to Implement the Free Trade Agreement between Canada and the United States of America) in the House of Commons in June 1988, the Liberal Party and the New Democratic Party, assisted by the Pro Canada Network, mounted a massive campaign to defeat the free trade deal.[1]

In an attempt to stop the legislation in its tracks, the Liberal Party used its majority in the Senate to force Bill C-130 to go through the Liberal-dominated Foreign Affairs Committee and subject it to an extensive series of hearings. At the same time, Allan MacEachen, the Liberal leader in the Senate, declared that the Senate would not approve the legislation unless Mulroney turned the issue over to a national vote. Faced with the intransigence of Liberal Senators, the Prime Minister dissolved Parliament and called a national election to be held in November 1988. When the electoral results came out, the Tories had garnered a majority of the seats in the Commons, winning 170 of 295 seats. Armed with a mandate to proceed with the enactment of the free trade legislation, the new Tory government resubmitted the free trade bill to the Parliament, where it was quickly enacted late in December 1988.

---

1. The Pro Canada Network (PCN) was an anti-free trade umbrella organization, civil society-based, composed of environmental, church, women, and native groups.

Conclusion

By capitalizing on certain state strengths originating from Canada's preexisting political economic institutions, policy entrepreneurs were able to implement liberal continentalism in a consistent and incremental manner. The strategy made full use of the federal state's market instruments already at its disposal and of its affinity with the ideology of limited government intervention. Moreover, the strategy and the arm's-length business-government relationship produced an ideal match between task and capacity—that is, it enhanced the autonomy of the private sector while it gradually reduced the scope of state authority. Finally, liberal continentalism capitalized on competitive federalism —the combination of which contributed, especially since the 1990s, to market-enforcing federalism.

Chapter Seven

Industrial Adjustment in the 1990s and Beyond

As Prime Minister Mulroney was on his way to visit Mexican President Carlos Salinas de Gortari in March 1990, he received news from Derek Burney, Canadian Ambassador to the United States at the time, that the Mexican president had earlier proposed to the United States that the two countries initiate bilateral free trade negotiations. When Mulroney was informed by Salinas himself of Mexico's free trade initiative, the Prime Minister left unanswered whether Canada would join. In this concluding chapter I will explain why Canada decided eventually to join Mexico and the United States to establish a trilateral free trade area. Moreover, the chapter will address how the strategy of liberal continentalism has shaped the Canadian economy in the 1990s and early 2000s.

The North American Free Trade Agreement: Defensive Positioning and Deepening Integration

From March 1990 to late September of that year, the Mulroney government carefully weighed the pros and cons of a trilateral free trade agreement. On September 24, the Prime Minister announced that Canada would participate in trilateral talks. Initially, support for a three-way trade agreement was thin, while the reasons for opposing it were numerous. One reason for opposing such an agreement was that the integration of the Canadian economy with a market as large and already interconnected as that of the United States would be more than enough to bring about the kind of industrial rationalization and specialization Canada was looking for. Mexico's smaller market and the limited trade between

Canada and Mexico initially convinced many in Canada that the economy would not benefit in terms of achieving more industrial transformation and economic growth if Canada were to opt to integrate its economy with that of Mexico's.

Second, the Canadian economy was still coping with the adjustment costs derived from the Canada-U.S. Free Trade Agreement (FTA), a situation made worse by the recession at the time. Although industrial restructuring did create new employment opportunities and led to business expansion, such developments were often eclipsed by news of plant closings and layoffs caused by industrial rationalization and the economic slowdown. It was believed that integration with Mexico, a low-cost producer, would bring about further market dislocations, particularly in the furniture, shoe, and garment industries, as investors relocated to Mexico to take advantage of lower production costs. Finally, the Tory government had gambled hugely when it decided that Canada would enter into a free trade agreement with the United States. Although it managed to marshal sufficient votes in support of the bilateral agreement, post-1988 public opinion trends began shifting against free trade. In a February 1991 opinion survey, 46% of Canadians supported a trilateral agreement, whereas 50% opposed it (Eden and Molot 1992, 75).

By late spring 1990, the debate was much more even-handed between those in favor and against; by late summer, a consensus in support of a trilateral agreement had emerged within the government and a pro-NAFTA coalition, involving the Tories, the BCNI, and most provinces with the exception of Ontario, Manitoba, and British Columbia, was formed. A number of factors contributed to this important shift. First, the government was concerned about protecting the relative gains it had secured through the FTA. Central to this concern was the belief that a bilateral Mexico-United States free trade agreement would undercut Canadian gains derived from its own bilateral arrangement with the United States. This would come about partly by improving Mexico's trade position relative to Canada and partly by enhancing the United States' position relative to both countries. The logic of this argument was furnished by two economists, Richard Lipsey (1990) and Ronald Wonnacott (1990), in their works on the hub-and-spoke model. According to the model, two separate bilateral free trade arrangements, both involving the United States, would confer on the United States (the hub) the benefit of free trade access to both Mexico and Canada (the two spokes), while the two spokes would not enjoy free trade access to each other's markets and would face the possibility that one could end up receiving more favorable free trade access to the United States than the other.

In terms of the distribution of gains, this model demonstrated how Canada could be at a disadvantage if it chose not to participate in negotiations. American exporters would enjoy duty-free access to both countries, whereas Canada and Mexico would only have such access to the United States. Thus, whatever economic losses Canada incurred as a result of heightened Mexican competition in the U.S. market could not be offset by an increase in Canadian trade to Mexico. In addition, Canada's losses could become greater if the bilateral Mexico-United States agreement provided Mexico with more preferential access to the United States than what Canada obtained under the FTA.

The model gained further attention when it was applied to investment patterns. In the absence of a deal that gave the two spokes reciprocal duty-free access to each other's markets, Canadian firms seeking to do more business in Mexico could decide to relocate to the United States, where they could enjoy such access to the Mexican market. The hub-and-spoke model advanced a scenario in which the United States would become the preferred location for investors seeking access to the two spokes and consequently drain Canada's industrial strength.

Accordingly, Canada's decision to join the NAFTA negotiations was profoundly shaped by what Grieco (1990) calls the 'defensive positionalist' rationale. In particular, Canada's overriding objective in the trilateral negotiations was to protect the relative gains it had secured through the bilateral trade arrangement. Canadian negotiators were particularly keen on resisting American attempts to undo or water down the FTA provisions that were beneficial to Canada. As Minister of International Trade Michael Wilson remarked in 1991, "Canada [was] not going to let the United States get through the back door [NAFTA] what it failed to get through the front door [FTA]" (quoted in Cameron and Tomlin 2000, 77).

The second factor contributing to Canada's decision to enter into trilateral talks was Ottawa's interest in improving the FTA. As Wilson (2000, 209) notes: "We were not just looking to protect what we had won in the FTA, we wanted to enhance the FTA." Canada wanted to add a new chapter on intellectual property rights and expand existing chapters on government procurement and financial services. Ottawa also sought to add greater precision and clarity to existing provisions in the FTA, particularly with respect to rules of origin and Chapter 19.

The Honda customs case illustrated that more precise regulations were needed to assist in the evaluation of rules of content. In March 1992, the U.S. Customs Service ruled that Canadian exports of Honda Civics from 1989 to 1990 failed to meet the FTA rules of content requirement. According to U.S. Customs, the American-made engines used in these Canadian assembled cars were manufactured with too many foreign parts. The ruling provoked a strong response in Canada because Revenue Canada had concluded that these engines conformed with the agreement's rules of origin standard when they were imported into Canada (International Trade Reporter 1992, 384). As for Article 1906 of Chapter 19, Canada and the United States had agreed that "the provisions of this Chapter [would] be in effect for five years pending the development of a substitute system of rules in both countries for antidumping and countervailing duties as applied to their bilateral trade." The provision went on to state that both countries would be given an additional two years to come up with a common set of rules, and if no agreement emerged at the end of this extension period, either country could choose to abrogate Chapter 19. Because this set of provisions was crucial for Canada, trilateral talks would give Ottawa an opportunity to fix the loophole tied to the provisional nature of Chapter 19.

Gaining access to the Mexican market was the least important of the three factors that persuaded Canada to participate in trilateral talks. Mexico was Canada's seventeenth most important trading partner in 1988, and its market had been protected by high trade barriers. However, Canada recognized that a trilateral agreement would open up the Mexican economy to Canadian products and in turn help Canada to diversify its trade pattern. In sum, Ottawa was going to sell NAFTA to Canadians by demonstrating how the new deal conserved and selectively improved Canada's position within a larger free trade area.

After an arduous eighteen months of negotiations, Ottawa was able to present a deal to Canadians that guaranteed the preservation of the distribution of gains it had secured under the FTA. Many of the agreement's provisions were folded into NAFTA, and the trilateral agreement encompassed new and improved terms that benefited Canada (Hart 2002, 395). With respect to improvements, Chapter 4 of NAFTA included precise language concerning the rules of content, Chapter 10 increased the liberalization of government procurement markets, and Chapter 17 required member countries, in particular Mexico, to adhere to international standards regarding protection of intellectual property. Moreover, unlike the investment chapter in the FTA that covered only foreign direct investment, Chapter 11 of NAFTA covered a broader range of investment activities, such as equity and debt securities, loans to firms, and business real estate, for example. Also, it established a dispute settlement mechanism that allowed investors to challenge the treatment it received in another NAFTA country.

Canadian and American negotiators clashed on three issues. Since coming into effect in 1989, the American Congress had become incensed with the dispute settlement mechanism established under Chapter 19 because it limited its ability to exert influence over the conduct of American trade policy. Thus, the American negotiating team, led by Carla Hills, fought hard to weaken Chapter 19 by making it easier for a country to question a decision of a binational panel by relaxing the extraordinary challenge procedure in Article 1904. Under this article, a country could invoke the extraordinary challenge procedure in which an Extraordinary Challenge Committee would be formed specifically to determine whether a binational panel abused its powers, violated procedures, or acted improperly.

By seeking to establish a special review mechanism in NAFTA, the American proposal, according to Canadian negotiators, would challenge the authority of binational panels. Canada effectively resisted such an effort and, after threatening to leave the negotiations, an agreement was struck that incorporated Chapter 19 of the FTA into NAFTA with minimal changes. Under the NAFTA's Chapter 19, the provision that allowed for the abrogation of the dispute mechanism once the seven-year period expired ("sunset termination") was removed, thus ensuring its permanency. Moreover, a number of new provisions were added that required Mexico to institute a trade remedy regime similar to that of Canada and the United States and to perform administrative practices of trade remedy laws similar to those in the other two countries.

Canada was also able to retain the right to screen foreign investment as outlined in the FTA. In the trilateral negotiations, Canadian negotiators were adamant about not giving up Canada's foreign investment screening authority. Canada's bargaining position was bolstered when Canadian negotiators managed to convince Mexican negotiators not to give up their foreign investment screening powers. Given Canada's intransigence and the fact that Mexico now wanted to retain its screening power, the United States had to drop its original demand of eliminating foreign investment screening. For Canada, this meant the incorporation of the investment screening rights secured under the FTA into Chapter 11 of NAFTA—namely, the right to review foreign takeovers that exceeded $150 million and review any acquisition in the oil, gas, and uranium sectors.

Finally, Canada was able to retain its right to protect the cultural industry—in particular, film and videos, musical recordings, radio and television broadcasts, and printed publications. Knowing very well that Canada would not yield to the American request that it open its cultural industry, American negotiators discontinued their challenge of the cultural industry exemption Canada had gained through Article 2005 of the FTA. The said exemption was incorporated into Chapter 21 of NAFTA.

NAFTA would intrude into provincial jurisdiction to a greater extent than did the FTA. As a result, the provinces were consulted and briefed regularly throughout the trilateral negotiations using an intergovernmental mechanism similar to the one used in the FTA negotiations. In total, six First Ministers' meetings were held, in addition to the numerous meetings and discussions among officials from both levels of government. Moreover, the two side deals of NAFTA, the North American Agreement on Environmental Cooperation and the North American Agreement on Labor Cooperation, affect the provinces far more than the federal government. Accordingly, the imperative of working closely with the provinces in order to formulate a Canadian proposal and ultimately to gain provincial compliance figured prominently in the way Ottawa proceeded in these negotiations. In no previous international negotiations had the provinces been so involved as in the negotiations leading to the finalization of the side deals (Skogstad 2002, 164-5).

The FTA and NAFTA have helped integrate Canada's internal market. The two regional free trade arrangements have caused some negative integration to the extent that they have required the provinces and Ottawa to limit their ability to employ particular policies that impede internal trade. Specifically, governments at the provincial and federal levels are prohibited from imposing restrictions on investment (Chapter 11 of NAFTA) and on financial services (Chapter 14 of NAFTA). The provinces are required to lower barriers on the distribution of wine and distilled spirits (Chapter 8 of FTA), and both levels of government are prohibited from imposing restrictions in the energy sector and from pursuing policies that interfere with the principle of national treatment.

What is unique about these coordinated efforts between the two levels of government is the fact that they have been the principal driving force behind the mutual withdrawal of federal and provincial states from the economy. Equally important is that these joint actions have operated alongside these governments' individual market-imposing efforts—a normal outcome when actors coexist under competitive federalism. On the one hand, the autonomy afforded to the provinces through competitive federalism has allowed them to pursue market liberalism in accordance with their interests, often following the same course of policy actions that have proven effective in other provinces. On the other hand, when both levels of government have recognized a common intergovernmental interest, it has led to coordinated efforts such as those witnessed in the negotiations of FTA and NAFTA.

While NAFTA was negotiated when the Tories were in office, it came down to Prime Minister Jean Chrétien's Liberal government to implement the agreement. After enjoying a comfortable majority for the past eight years, the parliamentary election in October 1993 shattered the Tory majority, reducing it to just two seats in the House of Commons. With the Liberals in power, many officials, remembering the party's anti-free trade position in the 1988 elections, believed that the new government would not enact NAFTA. On the contrary, the Liberals accepted free trade; in fact, they committed themselves to proceed with the strategy of market liberalism as the Tories had done. As Bradford (1998, 126) notes, Chrétien's Liberals "exhibited important continuities with [their] PC predecessors. The basic policy direction was unchanged: the steady embedding of a neo-liberal national economic policy discourse."

The institutionalization of market liberal policies was demonstrated with the ratification of NAFTA. It was also illustrated when the provinces and Ottawa signed the 1995 Agreement on Internal Trade (AIT). This reflected their commitment to achieve domestically what FTA and NAFTA had achieved in terms of deepening and broadening trade within the North American region. Although the AIT did not undermine provincial ability to act autonomously, the agreement did enhance negative integration in several ways. It eliminated some discriminatory procurement practices and most discriminatory pricing; it prevented governments from using residency requirements as a condition for occupational licensing and professional services; and it prevented governments from using various investment restrictions as well as certain types of industrial and agricultural subsidies. In addition to mutually reducing and eliminating barriers to interprovincial trade in goods and services and to the mobility of people, the AIT set up a dispute settlement procedure to adjudicate controversies arising from cases of nonconformity with the AIT. The AIT represented an important development in intergovernmental relations inasmuch as governments committed themselves to substituting market forces for political judgments in many areas of the Canadian internal market (Brown 2002).

Industrial Adjustment since the Mid-1990s

Canadians understand that their economic prosperity depends on keeping an open economy. But they also are especially aware of the fact that basic economic fundamentals—such as national innovation, productivity, growth rate, employment, and income—are all deeply influenced by Canada's ties with the United States. Since NAFTA came into effect, the Canada-U.S. economic relationship has grown closer and more complex. Through increased cross-border trade and investment, many Canadian industries are integrated into North American supply chains where they service both continental and international markets. What is more, cross-border rationalization and specialization of production has helped Canadian producers to move up the continental and global value chains, exporting commodities with high technology/high value-added content. In light of these developments, "economic integration or linkages between the two countries has reached the stage that Canada's economic well-being has become directly tied to an open and well-functioning bilateral border" (Hart and Dymond 2001, 10). With much of Canada's economic prosperity riding on continued access to the American market, it has been imperative for Canada to work closely with the United States in the post-September 11 bilateral environment to devise mutually beneficial arrangements that enhance border security while minimizing disruptions to cross-border commercial traffic. The following sections of the concluding chapter address the different themes introduced above.

Canada-United States Cross-Border Integration

Hart and Dymond (2006, 131) note that "the relationship with the United States is . . . the indispensable foundation of any Canadian policy to maximize benefits from engagement in the global economy." Indeed, Canada's ability to succeed in a globalized economy rests on the core competencies it has developed as a result of the acceleration and deepening of cross-border integration of the Canadian and American markets since the early 1990s. The two economies have never been so deeply integrated. On an average day in 2005, about $1.82 billion in goods and services traded across the border. In 2005, approximately 45,737 Canadian-based firms exported $401.5 billion worth of merchandise. Among them, 27,151 (59.3%) exported merely to the United States roughly $194.6 billion worth of merchandise; 7,488 (16.4%) exported only to non-U.S. destinations about $18.2 billion worth of goods; and another 11,098 (24.3%) exported to both about $188.7 billion worth of merchandise. Since 2001, there has been a decline in the number of Canadian-based firms that export exclusively to the United States, but an increase in the number of firms that export exclusively to non-U.S. markets and to both U.S. and non-U.S markets. Moreover, only 3.9% of all exporting firms accounted for 83% of the total value of exports (Statistics Canada 2007, 23-4).

The two-way trade figures clearly reveal how dependent Canada and the United States are on each other. In 2005, the Canadian share of U.S. merchandise imports was 16.9% ($291.9 billion), while the Canadian share of U.S. merchandise exports was 23.3% ($211.4 billion). The U.S. dollar value of merchandise imports from Canada exceeded those of any other country. The second and third largest suppliers of U.S. merchandise imports were China ($259.8 billion) and Mexico ($172.5 billion). Moreover, Canada received the largest amount of U.S. merchandise exports, followed by Mexico ($120.0 billion) and Japan ($55.4 billion).

The U.S. figures prominently in Canadian trade data. In 2005, U.S. share of Canadian merchandise exports and imports was 83.8% ($365.8 billion) and 56.5% ($215.2 billion), respectively. In the first half of the 2000s, the United States received the bulk of exports from key Canadian industries. Although export concentration in the United States dropped slightly, 97% of Canadian exports of motor vehicles and parts[1]—accounting for 21% of total exports to the United States—and 94.95% of exports of refined petroleum and coal products were destined for the United States.[2]

These trade figures, however, overlook two important facts about Canadian-U.S. cross-border integration. First, despite the widespread concern about whether the elevated U.S. share of Canadian exports may be creating vulnerability, a study by Goldfarb (2006) shows that such export concentration has not translated into lower and less stable average annual export growth for Canada. Even though Canada's level of export concentration has been the highest (with the exception of Mexico's) among OECD countries, its average annual export growth between 1991 and 2005 was 13%, just below the 15% average, and its level of variability in the growth rate was far lower than the average. In fact, in light of the trade-off between these two indicators, Canada's position has been relatively good. Many advanced industrialized countries with growth rates close to or above Canada's have witnessed comparatively much

---

1. The "motor vehicles and parts" reference combines such activities as the manufacturing of motor vehicles, motor vehicle bodies and trailers, and motor vehicle parts.

2. In 2005, the U.S. share of Canadian exports of key industries was as follows: machinery, 74.92%; primary metal products, 76.97%; transportation equipment, 93.92%; mining, oil and gas extraction, 85.78%; forestry and logging, 56.02%. Export concentration for these 7 industries dropped, on average, by 4.22 percentage points between 2000 and 2005. (Data retrieved from Canada, Industry Canada, Trade Data Online database, and provided by Statistics Canada).

higher export volatility, while those with lower volatility have seen slower growth. Moreover, had Ottawa actively pursued trade diversification since the mid-1990s, the evidence suggests that it would not have rendered Canada's economic prosperity more secure. The risk-reducing benefit of export diversification has its costs. As Goldfarb (2006, 18) observes, "Over the past decade, Canadian exports to the U.S. have been less volatile on average than have exports to most other regions. Shifting exports away from the U.S. over the past decade would likely have increased volatility and decreased trade growth, making Canada worse off."

Another important fact to bear in mind is the recent changes in the import content of Canadian exports. In their updated study on import content of exports in Canada, Cross and Ghanem (2008) calculate that import content in 2004 was 27.3%, down from 31.2% in 2000. The decline of the Canadian dollar in the early part of the decade led a majority of industries to cut their use of imports to produce exports. However, the slowdown in the use of imports was most notable in the automobile as well as in the machinery and equipment industries, which have maintained higher-than-average import contents in their exports. The rebound of the dollar beginning in 2003 resulted in a slight increase in 2004 in Canadian firms' use of imports in their production process. Furthermore, the upturn in import content coincided with the recovery in the share of value-added exports in GDP, which had dropped to 27.8% in 2003 from a high of 31.4% in 2000.

The supply chains into which key Canadian industries have been integrated continue to have profound continental profiles. The U.S. share of Canadian imports of intermediate inputs—which accounted for two-thirds of total U.S. exports to Canada—dropped from 70% in 1999 to roughly 62% in the mid-2000s. Still, as Cross and Ghanem (2008, 10) observe: "The U.S. dominance reflects the need for parts imported to feed domestic assembly lines to be close by, especially as manufacturers adopt just-in-time inventory systems. Very few could wait for the month-long voyage from China, or support the cost of air travel from Asia or Europe." And, even though Canadian exports to markets in Asia-Pacific and Europe have increased in recent years, the majority of its intermediate and end-point commodities supply the American market.

The rise in the import content of Canadian exports from the mid-1980s to 1999, followed by its relative stability in the first half of this decade, reflects the increasing specialization of many industries in Canada. Integrative trade with the United States and, increasingly, with other countries is central to Canada's continued economic strength.[3] As Blank (2005) observes: "Continentally

---

3. According to Hodgson (2004), integrative trade "encompasses all the emerging elements of international business – exports, imports used in exports, foreign investment and the use of foreign affiliates, especially for the sale of services in foreign markets."

integrated industries have become more efficient and more competitive"—the very industries that serve as engines of growth and innovation for Canada. Cameron and Cross (1999) point out that value added output per worker in the export sector exceeded that of the overall economy by about one-third in the 1990s. Moreover, "[t]he changing intensity and composition of bilateral trade," according to Hart and Dymond (2006, 131), "have contributed significantly to making Canadians better off both as consumers and as producers. Canadians employed in export-oriented sectors have consistently been better educated and better paid than the national average."

Cross-border investment and trade are inextricably tied together. In 2005, U.S. foreign direct investment (FDI) in Canada stood at $248.5 billion, which represented about 10% of all U.S. FDI abroad, but amounted to 62.9% of all inward FDI in Canada. The value of Canadian FDI abroad in 2005 was $455.2 billion, of which $202.7 billion (44.5%) was located in the United States. While the United States was Canada's most important destination of FDI, Canada was the United States' second-leading destination.

The presence of Canadian and American capital in each other's economy did not change significantly in the second half of the 1990s. While American capital has a greater effect on the performance of the Canadian economy than does Canadian capital on the U.S. economy, it is important to point out that the scale of American control of the Canadian economy was stable during the 1990s and was slightly lower in the early 2000s than in the early 1980s. There were 1,863 non-bank majority-owned Canadian affiliates of U.S. companies in 2000, down from 1,917 in 1995. While their share of Canada's GDP was 10% in 2000, up 1.4 percentage points from 1995, their share of Canadian employment rose from 6.3% in 1995 to 6.9% in 2000. The number of majority-owned American affiliates of Canadian companies dropped from 1,142 in 1995 to 843 in 2000. Despite the scale back, their contribution to the American economy increased— as evidenced by the rise in their share of the U.S. GDP (.33% in 1995 to .38% in 2000) and of American employment (.39% to .42%). Overall, the effect of FDI flows on the Canadian economy rose in the 1990s. Between 1991 and 1995, net FDI inflows equaled 1.1% of GDP and net FDI outflows equaled 1.3% of GDP. Between 1996 and 2000, FDI inflows and outflows were 4.0 and 4.1% of GDP, respectively.

A string of recent foreign takeovers of large Canadian companies have raised public concerns about the hollowing out of corporate Canada. Moreover, the potential sale of the Information Systems Business of Macdonald, Dettwiler and Associates Ltd. (MDA)—which manufactures space technologies such as satellites and the robotic arm used by the U.S. space shuttle and International Space Station—to the Minnesota-based Alliant Techsystems in early 2008, heightened fears that the United States has been snapping up Canada's prized technologies.[4] Indeed, the United States has been behind several of the major takeovers since 2000, but so have investors from other countries. For example, Dofasco, a steel producer, was bought by Arcelor of Luxembourg. The two mining giants, Falconbridge and Inco, were bought by Switzerland's Xstrata and Brazil's Vale, respectively. Alcan, the aluminum producer, was purchased by Rio Tinto, an Anglo-Australian outfit. The two hotel chains, Four Seasons and Fairmont, were bought by Saudi and American investors. Earlier, in 2000, Vivendi SA, a French company, bought Seagrams Co. Ltd. and British American Tobacco, a British firm, purchased the tobacco operations of Imasco Ltd.

For its part, corporate Canada has actively pursued investment opportunities abroad. In fact, since the mid-1990s, Canada has been a net acquirer of foreign assets. In 2004, the cumulative value amount of Canadian direct investment abroad exceeded that of foreign direct investment in Canada by 22% (Hejazi 2007, 3). Between 2001 and 2005, FDI inflow as a share of Canada's GDP averaged 2.2% (ranking it forty-sixth out of seventy-three advanced industrialized and developing countries), while FDI outflow averaged 3.8% (ranking Canada thirteenth among the seventy-three). During the same period, Canadian firms bought twenty-three foreign firms whose value exceeded $1 billion each, while foreign companies acquired twenty-five Canadian companies with assets greater than $1 billion each (Mintz and Tarasov 2007, 2-5).

The Canadian economy has attracted leading multinational corporations from around the world, while corporate Canada has expanded its outward reach. Global markets recognize Canada as a significant player in several global supply chains. Today, the main source of Canada's industrial strength and economic prosperity is traced to its industries' expanding role in continental- and global-oriented integrative trade. As a Conference Board of Canada Report (2006) points out: "Inward investment injects new technologies and know-how into [Canada's] economy. Canadian investment abroad creates and strengthens international supply chains and multiplies [Canada's] export potential." On the one hand, Canadian-owned and foreign-owned companies operating in Canada are increasingly involved in the process of integrating imports into the production of value-added exports. On the other hand, more and more Canadian-owned companies operating in Canada and abroad are manufacturing capital-intensive inputs that are incorporated into global products.

Post-9/11 Worries and Opportunities

Much of the economic adjustment that Canada has undertaken over the past two decades has been shaped by economic integration with the United States. So

---

4. In early May of 2008, Jim Prentice, the Minister of Industry, announced that ATK could not acquire MDA on the grounds that the transaction did not produce net benefits to Canada.

when the American government beefed up border security in the wake of the tragic events of September 11, 2001, Canadians began wondering what economic impact this measure would have on their economy. The common border plays an important role in both economies. It is estimated that 70% of bilateral trade is transported by truck; 30,000 trucks crossed the border in 2000; 15 million Canadians travel to the United States for a day or more each year in a typical year; and about 200 million individual crossings occur each year. Moreover, cross-border fragmentation of production has rendered industries in both countries vulnerable to sudden cross-border slowdowns. According to one estimate, delays in the arrival of auto parts can result in losses of $25,000 per minute for the automobile industry (Hart and Dymond 2001, 8). Another study calculates that the steel industry has suffered annual losses of US$300-$600 million due to delays in shipments (Pastor 2008, 88).

However, given the asymmetry in the importance of trade between the two countries, the economic costs associated with periodic border crossing slowdowns and other forms of disruption to cross-border flows are likely to be disproportionately larger for the Canadian economy. According to Dobson (2002, 16), Canadian productivity and investment environment could be severely undermined:

> The potential longer-term impact of a permanent increase in border security would be to raise transaction costs, acting like an added tariff on two-way trade. The higher costs of Canadian imports would cause U.S. customers to switch to cheaper domestic and alternative foreign suppliers. Higher costs of moving people, goods, and services would also undermine Canada's productivity performance. If international investors headed for North America bypass Canada, going into the United States to serve Canada from there, Canada's productivity growth would be negatively affected.

Thus, in the post-September 11 bilateral environment, the challenge facing Canada has been how to address both its vulnerability to market access disruptions and America's security concerns.

Here, McDougall's (2006, 310) "impossible trinity" formulation is instructive insofar as it helps to sort out different options for Canada. Only two of the following three objectives can be pursued at the same time: fulfill Canada's share of securing the North American perimeter so that Americans feel secure and remain supportive of further economic integration; create counterweights to American influence by strengthening economic ties with other regions and pursuing domestic policies that enhance Canada's economic sovereignty; or maintain fiscal discipline. The two general views that dominate the debate about Canada's options in a post-9/11 bilateral environment are both inspired by the imperative to preserve Canada's sovereignty and policy autonomy and to keep the country's fiscal state in order. However, whereas one approach calls for investing in the development of counterweights in a fiscally responsible way, the other calls for investing, in a fiscally sound way, in the common border security infrastructure to make it possible in the near future to build a North American economic space.

The first approach is based on the contention that Canada does not possess the leverage necessary to secure the level of market access it desires, and that the country will have to sacrifice too much sovereignty if it aligns itself with the United States in creating a comprehensive North American border security plan.[5] Winham and Ostry (2003) remark that the kind of economic security that Canada seeks from the United States could only be obtained by becoming the fifty-first state of the Union. Thus, a central economic policy recommendation stemming from this view is that Canada should "attempt to confront the U.S. export dependency that is at the root of the present uncertainties facing Ottawa. What is needed now is a major analytical effort to determine what scope exists in the Canadian economy, with permissible government support, to expand trade with regions outside North America, and especially to South America."

On the security front, proponents of this view lament the loss of sovereignty and the excessive financial costs resulting from Canada's role in securing the North American common perimeter. In expressing these sentiments, Byers (2007, 214) warns that Canada needs "to avoid the mistake of thinking that the U.S. approach [to protecting the common perimeter] must be right for Canada, too. When it comes to border security, the better approach is one of restraint."

The other view holds that through the adept exercise of Canadian sovereignty, Ottawa could pursue a strategic bargain with Washington in which Canada's firm commitment to helping the United States to secure the common perimeter is reciprocated by American willingness to consider different measures to enhance market access and deepen economic integration.[6] What the proponents of the 'big idea' approach, as it is commonly referred to, envision is "cooperation between neighbors to produce the public goods of homeland security and economic stability that neither country can produce on its own" (Dobson 2002, 5). But Canadians are aware that Americans have not assigned equal weight to each common objective; instead, security cooperation is conducive to economic cooperation. As Hart and Dymond (2001, 45-7) observe: "As a first step toward establishing a more open border, [Ottawa] will need to shore up U.S. confidence in Canada's ability to do its share in securing North America's common perimeter. If the government succeeds in rebuilding U.S. confidence, Canada will be better positioned to work with the United States and take the necessary steps to create confidence in a more secure common

---

5. This view has been most vividly expressed by Lloyd Axworthy, a former Minister of Foreign Affairs, and Michael Byers. See Axworthy (2005) and Byers (2007).
6. The phrase "exercise of Canadian sovereignty" is borrowed from Dobson (2002, 18). As she points out: "A nation that exercises its sovereignty anticipates change, prepares options that promote the key interests of its partner, but channels actions in ways that best serve its own interests."

perimeter and a more open joint border [that facilitates deeper economic integration]."

Advocates of this view envision Canada acting proactively—providing the kind of vision and leadership that can lead the two countries to achieve their priorities. Although security cooperation is paramount for Americans, proponents of this approach point out that the provision of continental security benefits both Canadians and Americans. Still, the economics of bilateral cooperation remains the core issue for Canada. In particular, they argue that the institutional framework embodied in NAFTA is inadequate to oversee cross-border commercial flows. According to Hart, "reducing the 'tyranny of small differences' in Canadian and American regulations that is frustrating businesses, adding costs, and stymieing the benefits of further economic integration" is urgently needed.

In addressing the failure of NAFTA's dispute settlement mechanism, Allan Gotlieb (2002), a former Canadian ambassador to the United States, asks: "Shouldn't we be building on NAFTA to create new rules, new tribunals, new institutions to secure our trade? Wouldn't this 'legal integration' be superior to ad hoc responses and largely ineffective lobbying to prevent harm from Congressional protectionist sorties? Are there not elements of a grand bargain to be struck, combining North American economic, defence and security arrangements within a common perimeter?" Thus, from a Canadian perspective, the main objectives of economic cooperation would be to reduce trade-impeding regulatory differences between the two countries, enhance dispute settlement mechanisms, and introduce certain customs union and common market features to facilitate trade, investment, and the movement of people.

Conclusion

Since the 1970s, no debate has influenced the political economy of industrial adjustment in Canada more than the one between integrationists and counter-integrationists. From the early 1970s to the early 1980s, the counter-integrationist forces influenced the selection of policy tools used to pursue economic adjustments. As a result of the emergence of a new constellation of domestic and external forces in the 1980s, the integrationists have since assumed the task of promoting economic adjustment.

In a post-9/11 environment, Canadians, once again, have a choice to make—deepening economic integration with the United States or developing counterweights. Two things are certain. First, if Canadians rally behind the counterweight option, the strategy will be heavily geared toward internationally-oriented adjustment. Governmental efforts aimed at restricting market processes failed in the 1970s and early 1980s, as they would today, because of the primacy of neoliberal economic ideas and the political-economic institutional framework that supports them. The other reality is that neither option is without its political and economic costs. The integration option promises continued economic prosperity, but may lead to further erosion of Canada's political autonomy. The counterweight option guarantees to diminish Canada's vulnerability resulting from its dependence on the United States, but may jeopardize its future economic stability and growth.

References

Abonyi, Arpad.1988. "Government Participation in Investment Development." In *Canada Under Mulroney*, ed. Andrew Gollner and Daniel Salée, 158-85. Montreal: Véhicle Press.

Aitken, Hugh G. J. 1961. *American Capital and Canadian Resources*. Cambridge, MA: Harvard University Press.

———. 1964. "Government and Business in Canada: An Interpretation." *Business History Review* 38 (1): 4-21.

Alberta, Department of Industry and Commerce. 1974. "Management of Growth." Government of Alberta, May 29.

Archer, Keith, Roger Gibbins, Rainer Knopff, and Leslie Pal. 1999. *Parameters of Power: Canada's Political Institutions*. 2nd ed. Scarborough, ON: International Thomson Publishing.

Aktinson, Michael. 1985. "If You Can't Beat Them: World Product Mandating and Canadian Industrial Policy." In *Canada and the New International Division of Labour*, ed. Duncan Cameron and Francois Houle, 125-44. Ottawa: University of Ottawa Press.

Atkinson, Michael, and William Coleman. 1988. *The State, Business and Industrial Change in Canada*. Toronto: University of Toronto Press.

Aucoin, Peter. 1986. "Organizational Change in the Machinery of Canadian Government: From Rational Management to Brokerage Politics." *Canadian Journal of Political Science* 19 (1): 3-27.

Axworthy, Thomas. 2000. "To Stand Not So High Perhaps but Always Alone: The Foreign Policy of Pierre Elliot Trudeau." In *Towards a Just Society: The Trudeau Years*, ed. Thomas Axworthy and Pierre Elliott Trudeau, 56-92. Toronto: Penguin Books.

Azzi, Stephen. 1999. *Walter Gordon & the Rise of Canadian Nationalism*. Montreal: McGill-Queen's University Press.

Baldwin, John R., and Wulong Gu. 2005. "Global Links: Multinationals, Foreign Ownership and Productivity Growth in Canadian Manufacturing." *Statistics Canada Research Paper* Catalogue no. 11-622-MIE, December.

Barlow, Maude. 1990. *Parcel of Rogues: How Free Trade is Failing Canada*. Toronto: Key Porter Books.

Barrows, David, and Mark Boudreau. 1987. "The Evolving Role of the Provinces in International Trade Negotiations." In *Knocking on the Back Door*, ed. Allan Maslove and Stanley Winer. Halifax: Institute for Research on Public Policy.

Barry, Donald. 1984. "The US and Canada: the right balance." *Baltimore Sun*, September 21.

Bashevkin, Sylvia B. 1991. *True Patriot Love: The Politics of Canadian Nationalism*. Toronto: University of Toronto Press.

Bickerton, James. 1999. "Regionalism in Canada." In *Canadian Politics*, ed. James Bickerton and Alain-G Gagnon, 209-38. 3rd ed. Peterborough, ON: Broadview Press.

Black, Edwin, and Alan Cairns. 1966. "A Different Perspective on Canadian Federalism." *Canadian Public Administration* 9 (1): 27-45.

Blair, Robert. 1985. *Minutes of Proceedings and Evidence of the Standing Committee on Regional Development*. 33rd Parliament of Canada. Ottawa, February 27.

Blank, Stephen. 2005. "Three Possible NAFTA Scenarios: It is Time for Canada to Think Carefully about North America." *Embassy*, September 7.

Bliss, Michael. 1970. "Canadianizing American Business: The Roots of the Branch Plant." In *Close the 49th Parallel, Etc.: The Americanization of Canada*, ed. Ian Lumsden, 26-42. Toronto: University of Toronto.

————. 1974. *A Living Profit.* Toronto: McClelland and Stewart.

————. 1982. *The Evolution of Industrial Policies in Canada: A Historical Survey.* Ottawa: Economic Council of Canada.

Bothwell, Robert. 1977. "The Canadian Connection: Canada and Europe." In *Foremost Nation: Canadian Foreign Policy and a Changing World*, ed. Norman Hillmer and Garth Stevenson, 24-36. Toronto: McClelland and Stewart.

Bradford, Neil. 1998. *Commissioning Ideas: Canadian National Policy Innovation in Comparative Perspective.* Toronto: Oxford University Press.

Britton, John N.H., and James M. Gilmour. 1978. *The Weakest Link: A Technological Perspective on Canadian Industrial Underdevelopment.* Ottawa: Minister of Supply and Services Canada.

Brodie, Janine, and Jane Benson. 1988. *Crisis, Challenge and Change: Party and Class in Canada.* Ottawa: Carleton University Press.

Brooks, Stephen. 1983. "The State as Entrepreneur: From CDC to CDIC." *Canadian Public Administration* 26: 525-43.

Brooks, Stephen G. 1997. "Dueling Realisms." *International Organization* 51 (3): 445-77.

Brown, Douglas. 1988. "The Federal-Provincial Consultation Process." In *Canada: The State of the Federation, 1987-88*, ed. P.M. Leslie and R.C. Watts, 77-93. Kingston, ON: Institute of Intergovernmental Relations, Queen's University.

————. 2002. *Market Rules: Economic Union Reform and Intergovernmental Policy-Making in Australia and Canada.* Montreal and Kingston: McGill-Queen's University Press.

Burney, Derek. 2000. "Where There's the Will...." In *Free Trade: Risks & Rewards*, ed. L. Ian MacDonald, 61-9. Montreal and Kingston: McGill-Queen's University Press.

Byers, Michael. 2007. *Intent for a Nation: What is Canada For?* Vancouver: Douglas & McIntyre.

Cameron, G and P. Cross. 1999. "The Importance of Exports to GDP and Jobs." *Canadian Economic Observer* (November): 1-5.

Cameron, Maxwell, and Brian Tomlin. 2000. *The Making of NAFTA: How the Deal Was Done.* Ithaca, NY: Cornell University Press.

Cameron, Richard. 1998. "Intrafirm Trade of Canadian-Based Foreign Transnational Companies." *Industry Canada Working Paper* 26, December.

Campbell, Colin. 1983. *Governments under Stress: Political executives and key bureaucracies in Washington, London, and Ottawa.* Toronto: University of Toronto Press.

Canada. 1972. *Foreign Investment in Canada.* Gray Report. Ottawa: Information Canada.

————. 1973. *Canadian Relations with the European Community.* Ottawa: Queen's Printer of Canada.

————. 1977. *Review of Developments in the GATT Multilateral Trade Negotiations in Geneva.* Ottawa: Minister of Supply and Services.

————. 1980. Speech from the Throne. *Commons Debate*, April 14.

————. 1984. Speech from the Throne, *Commons Debate*, November 5.

————. 1981. *Economic Development for Canada in the 1980s.* Ottawa: Minister of Supply and Services, November.

————. 1987. *Investment Canada, 1987-1988, Estimates Part III.* Ottawa: Minister of Supply and Services Canada.

Canada, Board of Economic Ministers. 1979. *Action for Industrial Growth: Continuing the Dialogue.* Ottawa: Minister of Supply and Services, February.

Canada, Canada Development Investment Corporation. 1983. *Annual Report.* Ottawa: Supply and Services Canada.

Canada, Department of External Affairs. 1970. *Foreign Policy for Canadians.* Ottawa: Queen's Printer.

———. 1972. "Statement by External Affairs Minister Mitchell Sharp to the Conference on Canada and the European Community in Ottawa," November 2.

———. 1972. "An Address by the Secretary of State for External Affairs, the Honourable Mitchell Sharp, to the Buffalo Area Chamber of Commerce, Buffalo New York." *Statements and Speeches* 72/7, May 9.

———. 1975. "The Contractual Link—A Canadian Contribution to the Vocabulary of Cooperation." Remarks by Prime Minister Pierre Elliot Trudeau at the Mansion House, London, England. *Statements and Speeches* 75/6, March 13.

———. 1976. "The Contractual Link: Why and How? Address by Mr. Marcel Cadieux, Ambassador of Canada to the European Communities, to the Canadian Institute of International Affairs, Toronto." *Statements and Speeches* 76/33, November 24.

———. 1983. *Review of Canadian Trade Policy: A Background Document to Canadian Trade Policy for the 1980s.* Ottawa: Supply and Services Canada.

———. 1983. *Canadian Trade Policy for the 1980s: A Discussion Paper,* August.

———. 1984. "New Climate for Investment in Canada: Notes for a Speech by the Right Honourable Brian Mulroney, Prime Minister, to the members of the Economic Club of New York, New York." *Statements and Speeches* 84/18, December 10.

———. 1985. *How to Secure and Enhance Canadian Access to Export Markets,* January.

———. 1986. *Canadian Trade Negotiations: Introduction, Selected Document, Further Reading.* Ottawa: Minister of Supply and Services Canada.

Canada, Department of Finance. 1984. "A New Direction for Canada: An Agenda for Economic Renewal." Ottawa, November 8.

———. 1985. "The Budget Speech by Michael Wilson." Ottawa, May 23.

Canada, Department of Foreign Affairs and International Trade. 1996. *Canada: Drawers of Water, Hewers of Wood and Dangers of Other Myths.* Trade and Economic Policy Paper 96/07. Ottawa, December.

Canada, Department of Industry, Trade and Commerce. 1978. *Annual Report, 1977-1978.* Ottawa: Supply and Services Canada.

Canada, Economic Council of Canada. 1975. *Looking Outward: A New Trade Strategy for Canada.* Ottawa: Information Canada.

Canada, Energy, Mines and Resources Canada. 1980. *The National Energy Program, 1980.* Ottawa: Minister of Supply and Services Canada.

———. 1982. *The National Energy Program: Update 1982.* Ottawa: Minister of Supply and Services Canada.

Canada, Export Promotion Review Committee. 1978. *Strengthening Canada Abroad.* The Hatch Report, November 30.

Canada, Foreign Investment Review Agency. 1976. *Foreign Investment Review Agency: Annual Report, 1975-76.* Ottawa: Minister of Supply and Services.

———. 1979. *Foreign Investment Review Agency: Annual Report, 1978-79.* Ottawa: Supply and Services.

Canada, Investment Canada. 1987. *Investment Canada, 1987-1988 Estimates Part III.* Ottawa: Minister of Supply and Services Canada.

Canada, Major Projects Task Force (Blair-Carr Task Force). 1981. *Major Canadian Projects: Major Canadian Opportunities.* Report of the Major Projects Task Force. Ottawa: Supply and Services Canada.

Canada, Ministry of State for Economic Development. 1980. *Annual Report, 1979-1980.* Ottawa: Supply and Services Canada.

Canada, Office of the Minister of State (Privatization). 1987. "Excerpts from Statements made by the Honourable Barbara McDougall on the Reasons for Privatization," May.

Canada, Privy Council Office. 1968. *Foreign Ownership and the Structure of Canadian Industry: Report of the Task Force on the Structure of Canadian Industry.* Watkins Report. Ottawa: Queen's Printer.

————. 1977. *Crown Corporations: Direction, Control, and Accountability.* Ottawa: Minister of Supply and Services.

Canada, Royal Commission on Canada's Economic Prospects (Gordon Commission). 1957. *Report of the Royal Commission on Canada's Economic Prospects.* Ottawa: Queen's Printer.

Canada, Royal Commission on Economic Union and Development Prospects for Canada (Macdonald Commission). 1985. *Report of the Royal Commission on the Economic Union and Development Prospects for Canada.* 3 Vols. Ottawa: Minister of Supply and Services.

Canada, Science Council of Canada. 1976-7. "Technological Sovereignty: A Strategy for Canada." *Eleventh Annual Report.* Ottawa: Minister of Supply and Services Canada.

Canada. 1970. *Eleventh Report of the Standing Committee on External Affairs and National Defence Respecting Canada-U.S. Relations* (Wahn Report), Second Session, 28th Parliament. Ottawa: Queen's Printer.

Canada, Standing Committee on Finance, Trade and Economic Affairs. 1973. Statement of the Honourable Richard Hatfield Concerning Bill C-132 Foreign Investment Review Act. 29th Parliament of Canada. Ottawa: House of Commons, June 12.

————. 1973. Brief Represented by the Government of Saskatchewan Concerning Bill C-132, Foreign Investment Review Act. 29th Parliament of Canada. Ottawa: House of Commons, June 19.

————. 1973. Brief Submitted by the Committee for an Independent Canada Concerning Bill C-132, Foreign Investment Review Act. 29th Parliament of Canada. Ottawa: House of Commons, June 21.

————. 1973. Brief Presented by Government of Ontario. Re Bill C-132: the Foreign Investment Review Bill. 29th Parliament of Canada. Ottawa: House of Commons, July 19.

————. 1973. Brief Submitted by Executive Council of the Canadian Chamber of Commerce to the Honourable A.W. Gillespie, Minister of Industry, Trade and Commerce. Ottawa: House of Commons, July 19.

Canada, Standing Senate Committee on Foreign Affairs. 1973. *Canadian Relations with the European Community.* Ottawa: Queen's Printer for Canada.

Canada, Treasury Board. 1985. *Annual Report to Parliament on Crown Corporations and Other Corporate Interests of Canada.* Ottawa: Supply and Services Canada.

Chandler, William. 1987. "Federalism and Political Parties." In *Federalism and the Role of the State,* ed. Herman Bakvis and William Chandler, 149-70. Toronto: University of Toronto Press.

Chandler, William, and Herman Bakvis. 1989. "Federalism and the Strong-State/Weak-State Conundrum: Canadian Economic Policymaking in Comparative Perspective." *Publius: The Journal of Federalism* 19: 59-77.

Chow, Frank. 1993. "Recent Trends in Canadian Direct Investment Abroad: The Rise of Canadian Multinationals, 1969-1992." *Statistic Canada Research Paper* 8 Catalogue no. 11-010, December.

Clark, Joe. 1984. "Speech to the Strategic Planning Forum." Ottawa (October 25).

Clark-Jones, Melissa. 1987. *A Staple State: Canadian industrial resources in Cold War*. Toronto: University of Toronto.

Clarkson, Stephen. 1982. *Canada and the Reagan Challenge: Crisis in the Canadian-American Relationship*. Toronto: James Lorimer & Company, Publishers.

Clement, Wallace. 1977. *Continental Corporate Power: Economic Linkages between Canada and the United States*. Toronto: McClelland and Stewart.

Clifford, Wayne. 1991. "An Alberta Perspective." In *Canadian Federalism: Meeting Global Economic Challenges*, ed. Douglas Brown and Murray Smith, 22-45. Kingston, ON: Institute of Intergovernmental Relations.

Cody, Howard. 1977. "The Evolution of Federal-Provincial Relations in Canada: Some Reflections." *The American Review of Canadian Studies* 7 (1): 55-83.

Coleman, William D. 1985. "Analyzing the associative action of business: policy advocacy and policy participation." *Canadian Public Administration* 28 (3): 413-33.

Conference Board of Canada. 2006. "The Phantom Fears of Foreign Direct Investment." *Inside Edge* (Spring): 7.

Cross, P, and Z. Ghanem. 2008. "Tracking valued-added trade: Examining global inputs to exports." *Canadian Economic Observer* (February): 3.1-3.12.

D'Aquino, Thomas. 1985. *Minutes of Proceedings and Evidence of the Standing Committee on Regional Development*. 33rd Parliament of Canada. Ottawa, March 5.

De Boer, Stephen. 2002. "Canadian Provinces, US States and North American Integration: Bench Warmers or Key Players." *Choices* 8: 1-24.

Deeg, Richard. 2005. "Change from Within: German and Italian Finance in the 1990s." In *Beyond Continuity: Institutional Change in Advanced Political Economies*, ed. Wolfgang Streeck and Kathleen Thelen, 169-202. New York: Oxford University Press.

Dimma, William. 1974. "The Canada Development Corporation." Ph.D. Dissertation. Cambridge, MA: Harvard University.

Dobell, Peter. 1985. *Canada in World Affairs, Volume XVII, 1971-1973*. Toronto: Canadian Institute of International Affairs.

Dobson, Wendy. 2002. "Shaping the Future of the North American Economic Space." *C.D. Howe Institute Commentary* 162 (April): 1-32.

Doern, G. Bruce. 1982. "Liberal Economic Development Statement: To Scheme Virtuously." *The Canadian Business Review* (Spring): 48-53.

———. 1983. "The mega-project episode and the formulation of Canadian economic development policy." *Canadian Public Administration* 26: 219-238.

———. 1990. "The Department of Industry, Science and Technology: Is There Industrial Policy After Free Trade." In *How Ottawa Spends, 1990-1991*, ed. Katherine Graham, 49-72. Ottawa: Carleton University Press.

Doern, G. Bruce, and Brian W. Tomlin. 1991. *Faith and Fear: The Free Trade Story*. Toronto: Stoddart Publishing Co. Limited.

Dolan, Michael. 1978. "Western Europe as a Counterweight: An Analysis of Canadian-European Foreign Policy Behaviour in the Post-War Era." In *Canada's Foreign Policy: Analysis and Trends*, ed. Brian Tomlin, 26-48. Toronto: Methuen.

Doran, Charles. 1982-3. "The United States and Canada: intervulnerability and interdependence." *International Journal* 38 (1): 128-46.

Eden, Lorraine, and Maureen Appel Molot. 1993. "Canada's National Policies: Reflections on 125 Years." *Canadian Journal of Political Science* 19 (3): 232-51.

Evans, John. 1971. *The Kennedy Round and American Trade Policy*. Cambridge, MA: Harvard University Press.

Feigenbaum, Harvey, Jeffrey Henig, and Chris Hamnett. 1998. *Shrinking the State: The Political Underpinnings of Privatization.* Cambridge, U.K.: Cambridge University Press.

Finlayson, Jack, and Stefano Bertasi. 1992. "Evolution of Canadian Postwar International Trade Policy." In *Canadian Foreign Policy and International Economic Regimes,* ed. Claire A. Cutler and Mark W. Zacher, 19-46. Vancouver: University of British Columbia Press.

Fossum, John Erik. 1997. *Oil, the State, and Federalism: The Rise and Demise of Petro-Canada as a Statist Impulse.* Toronto: University of Toronto Press.

Foster, Peter. 1983. "Ottawa's Man: The Interventionist Politics of Maurice Strong." *Saturday Night* (August): 17-23.

Franck, Thomas, and K. Scott Gudgeon. 1975. "Canada's Foreign Investment Control Experiment: The Law, the Context and the Practice." *New York University Law Review* 50: 76-146.

French, Richard. 1979. "The Privy Council Office: Support for Cabinet Decision-Making." In *The Canadian Political Process,* ed. Richard Schultz, Orest Kruhlak, and John Terry, 2nd ed. Toronto: Holt, Rinehart and Winston.

———. 1980. *How Ottawa Decides: Planning and Industrial Policy-Making, 1968-1980.* Toronto: Lorimer.

Gerschenkron, Alexander. 1962. *Economic Backwardness in Historical Perspectives: A Book of Essays.* Cambridge: Belknap Press of Harvard University Press.

Gillies, James. 1981. *Where Business Fails: Business Government Relations at the Federal Level in Canada.* Montreal: Institute for Research on Public Policy.

Gillespie, Alastair. 1973. Statement before the House of Commons. *Commons Debate.* Ottawa: House of Commons, March 30.

———. 1973. Statement before the Standing Committee on Finance, Trade and Economic Affairs of the House of Commons. 29th Parliament of Canada. Ottawa: House of Common, June 5.

Gilpin, Robert. 1975. *U.S. Power and the Multilateral Corporation: The Political Economy of Foreign Direct Investment.* New York: Basic Books, Inc., Publishers.

———. 2001. *Global Political Economy: Understanding the International Economic Order.* Princeton, NJ: Princeton University Press.

Globe and Mail. 1971. "The beast is Loose." December 1.

———. 1984. "Business links Investment Canada." December 8.

———. 1985. "Carney sees energy prices falling." March 29.

———. 1991. "Bombardier acquires UTDC in Subsidy Deal with Ontario." December 5.

———. 1992. "Deal Moves Bombardier into Aerospace Elite." November 24.

Godfrey, Dave, and Mel Watkins. 1970. *Gordon to Watkins to You. Documentary: The Battle for Control of our Economy.* Toronto: New Press.

Goldfarb, Danielle. 2006. "Too Many Eggs in One Basket? Evaluating Canada's Need to Diversify Trade." *C.D. Howe Institute Commentary* 236 (July): 1-28.

Gotlieb, Allan. 2002. "Why not a grand bargain with the U.S.?" *National Post,* September 11.

Gourevitch, Peter. 1986. *Politics in Hard Times: Comparative Responses to International Economic Crises.* Ithaca, NY: Cornell University Press.

Granatstein, J.L. 1998. *The Ottawa Men: The Civil Service Mandarins, 1935-1957.* Toronto: University of Toronto Press.

Granatstein, J.L., and Robert Bothwell. 1991. *Pirouette: Pierre Trudeau and Canadian Foreign Policy.* Toronto: University of Toronto Press.

Gray, Herb. 1980. "Economic Nationalism and Industrial Strategies." Notes for an address to the Annual Symposium. École des Hautes Études Commerciales. Montreal, June 3.

———. 1981. "Canadian National Energy Forum: Energy and Industrial Development in Canada." Canadian National Committee World Energy Conference. Ottawa, November 9-10.

Grieco, Joseph. 1990. *Cooperation Among Nations: Europe, American, and Non-Tariff Barriers to Trade.* Ithaca, NY: Cornell University Press.

———. 1996. "State Interests and Institutional Rule Trajectories: A Neorealist Interpretation of the Maastricht Treaty and European Economic and Monetary Union." *Security Studies* 5 (3): 261-305.

———. 1997. "Systemic Sources of Variation in Regional Institutionalization in Western Europe, East Asia, and the Americas." In *The Political Economy of Regionalism*, ed. Edward Mansfield and Helen Milner, 164-87. New York: Columbia University Press.

Guillemette, Yvan and Jack M. Mintz. 2004. "A Capital Story: Exploring the Myths Around Foreign Investment in Canada." *C.D. Howe Institute Commentary* 201 (August): 1-32.

Hacker, Jacob S. 2002. *The Divided Welfare State: The Battle over Public and Private Social Benefits in the United States.* New York: Cambridge University Press.

Hale, Geoffrey. 2006. *Uneasy Partnership: The Politics of Business and Government in Canada.* Peterborough, ON: Broadview Press.

Hall, Peter. 1986. *Governing the Economy: The Politics of State Intervention in Britain and France.* New York: Oxford University Press.

———. 1993. "Policy Paradigms, Social Learning, and the State." *Comparative Politics* 25 (3): 275-96.

Hardin, Hershel. 1974. *A Nation Unaware: The Canadian Economic Culture.* Vancouver: J.J. Douglass Ltd.

Hart, Michael. 1990. *A North American Free Trade Agreement: The Strategic Implications for Canada.* Ottawa and Halifax: Centre for Trade Policy and Law and Institute for Research on Public Policy.

———. 1994. *Decision at Midnight: Inside the Canada-US Free-Trade Negotiation.* Vancouver: University of British Columbia.

———. 2002. *A Trading Nation: Canadian Trade Policy from Colonialism to Globalization.* Vancouver: University of British Columbia.

———. 2006. "Steer or Drift? Taking Charge of Canada-US Regulatory Convergence." *C.D. Howe Institute Commentary* 229 (March): 1-30.

Hart, Michael and Bill Dymond. 2001. "Common Borders, Shared Destinies: Canada, the United States and Deepening Integration." Ottawa: Centre for Trade Policy and Law.

———. 2006. "Trade Theory, Trade Policy, and Cross-Border Integration." In *Trade Policy Research 2006*, ed. Dan Ciuriak, 103-58. Ottawa: Minister of Public Works and Government Services Canada.

Heclo, Hugh. 1974. *Modern Social Policies in Britain and Sweden.* New Haven, CT: Yale University Press.

Hejazi, Walid. 2007. "Offshore Financial Centers and the Canadian Economy." University of Toronto. Mimeographed, February.

Houle, François.1987. "L'état canadien et le capitalisme mondial: Stratégies d'insertion." *Canadian Journal of Political Science* 20 (3): 467-500.

Hurtig, Mel. 1991. *The Betrayal of Canada.* Toronto: Stoddart.

Ikenberry, G. John. 1986. "The irony of state strength: comparative responses to the oil shocks in the 1970s." *International Organization* 40 (1): 105-37.

———. 1986. "The State and Strategies of International Adjustment." *World Politics* 39 (1): 53-77.

———. 1988. *Reasons of State: Oil Politics and the Capacities of American Government.* Ithaca, NY: Cornell University Press.

Inwood, Gregory J. 2005. *Continentalizing Canada: The Politics and Legacy of the Macdonald Royal Commission.* Toronto: University of Toronto Press.

International Trade Reporter. 1992. "Customs Rules that Canadian Honda Civics Failed to Meet Content Standards under FTA," March 4.

Jackson, Robert, and Doreen Jackson. 2001. *Politics in Canada: Culture, Institutions, Behaviour and Public Policy,* 5th ed. Toronto: Prentice-Hall.

Jenkins, Barbara. 1992. *The Paradox of Continental Production: National Investment Policies in North America.* Ithaca, NY: Cornell University Press.

Jenkins, Michael. 1983. *The Challenge of Diversity: Industrial Policy in the Canadian Federation.* Ottawa: Science Council of Canada.

Katzenstein, Peter. 1978. *Between Power and Plenty: Foreign Economic Policies of Advanced Industrial States.* Madison: University of Wisconsin Press.

Keating, Tom. 1993. *Canada and World Order: The Multilateralist Tradition in Canadian Foreign Policy.* Toronto: McClelland and Stewart Inc.

Keenes, Ernie. 1992. "Rearranging the deck chairs: A political economy approach to foreign policy management in Canada." *Canadian Public Administration* 35 (3): 381-401.

Keenleyside, Terence, Lawrence LeDuc, and J. Alex Murray. 1976. "Public Opinion and Canada-United States Economic Relations." *Behind the Headlines* 35 (4): 1-26.

Kelly, R. F. 1976. *Shareholder Survey.* Vancouver.

Kinsman, Jeremy. 1973. "Pursuing the realistic goal of closer Canada-EEC links." *International Perspective* (January/February): 22-7.

Knubley, John. 1987. *The Origins of Government Enterprise in Canada.* Ottawa: Economic Council of Canada.

Krasner, Stephen. 1985. *Structural Conflict: The Third World Against Global Liberalism.* Berkeley: University of California Press.

Kuznets, Simon. 1966. *Modern Economic Growth: Rate, Structure, and Spread.* New Haven, CT: Yale University Press.

Langdon, Frank. 1980. "Problems of Canada-Japan Economic Diplomacy in the 1960s and 1970s: The Third Option." In *Canadian Perspectives on Economic Relations with Japan,* ed. Keith Hay, 73-89. Montreal: The Institute for Research on Public Policy.

Langille, David. 1987. "The Business Council on National Issues and the Canadian State." *Studies in Political Economy* 24: 41-85.

Laux, Jeanne Kirk. 1993. "How Private is Privatization." *Canadian Public Policy* 19 (4): 398-411.

Laux, Jeanne Kirk, and Maureen Appel Molot. 1988. *State Capitalism: Public Enterprise in Canada.* Ithaca, NY: Cornell University Press.

Laxer, Gordon. 1989. *Open for Business: The Roots of Foreign Ownership in Canada.* Toronto: Oxford University Press.

LeDuc, Lawrence, and J. Alex Murray.1983. "A Resurgence of Canadian Nationalism: Attitudes and Policy in the 1980s." In *Political Support in Canada: The Crisis Years,* ed. Allan Kornberg and Harold Clarke, 270-90. Durham: Duke University Press.

————. 1989. "Open for Business? Foreign Investment and Trade Issues in Canada." In *Economic Decline and Political Change*, ed. Harold Clarke, Marianne Stewart, and Gary Zuk, 127-39. Pittsburgh, PA: University of Pittsburgh Press.

Leslie, Peter. 1987. *Federal State, National Economy*. Toronto: University of Toronto Press.

Levac, Mylene, and Philip Wooldridge. 1997. "The Fiscal Impact of Privatization in Canada." *Bank of Canada Review* (Summer): 25-40.

Leyton-Brown, David. 1974. "The Multinational Enterprise and Conflict in Canadian American Relations." *International Organization* 28 (Autumn): 733-54.

————. 1980-1. "Extraterritoriality in Canadian-American relations." *International Journal* 36 (1): 185-207.

————. 1986. "Canada-U.S. Relations: Towards a Closer Relationship." In *Canada Among Nations, 1985: The Conservative Agenda*, ed. Maureen Appel Molot and Brian Tomlin, 177-95. Toronto: James Lorimer.

————. 1988. "Canada-U.S. Trade Disputes and the Free Trade Deal." In *Canada Among Nations, 1987*, ed. Maureen Appel Molot and Brian Tomlin. Toronto: James Lorimer and Company.

Lipset, Richard. 1990. "Canada at the U.S.-Mexico Free Trade Dance: Wallflower or Partner." *C.D. Howe Institute Commentary* 20.

————. 2000. "The Canada-U.S. FTA: Real Results versus Unreal Expectations. In *Free Trade: Risks & Rewards*, ed. L. Ian MacDonald, 99-106. Montreal and Kingston: McGill-Queen's University Press.

Mace, Gordon, and Gerard Hervouet. 1989. "Canada's Third Option: A Complete Failure?" *Canadian Public Policy* 15 (4): 387-404.

Mahant, E.E. 1981. "Canada and the European Community: The First Twenty Years." *Journal of European Integration* 4 (3): 263-79.

Mastanduno, Michael, David Lake, and G. John Ikenberry. 1989. "Toward a Realist Theory of State Action." *International Studies Quarterly* 33 (4): 457-74.

McDougall, John N. 2006. *Drifting Together: The Political Economy of Canada-US Integration*. Peterborough, ON: Broadview Press.

Meadwell, Hudson, and Pierre Martin. 1996. "Economic integration and the politics of independence." *Nations and Nationalism* 2 (1): 67-87.

Mearsheimer, John. 1994-95. "The False Promise of International Institutions." *International Security* 19 (3): 5-49.

Michelmann, Hans. 1986. "Federalism and international relations in Canada and the Federal Republic of Germany." *International Journal* 41 (3): 539-571.

Milne, David. 1986. *Tug of War: Ottawa and the Provinces Under Trudeau and Mulroney*. Toronto: James Lorimer & Company, Publishers.

Mintz, Jack M. and Andrey Tarasov. 2007. "Canada is Missing Out on Global Capital Market Integration." *C.D. Howe Institute E-brief*, August 21.

Molot, Maureen Appel. 1988. "The Provinces and Privatization: Are the Provinces Really Getting Out of Business?" In *Privatization, Public Policy and Public Corporation*, ed. Allan Tupper and G. Bruce Doern, 399-425. Halifax: The Institute for Research on Public Policy.

————. 2005. "The Trade-Security Nexus: The New Reality in Canada-U.S. Economic Integration." *The American Review of Canadian Studies* (Spring): 27-62.

Molot, Maureen Appel, and Glen Williams. 1984. "The Political Economy of Continentalism." In *Canadian Politics in the 1980s*, ed. Michael Whittington and Glen Williams, 81-104. Toronto: Methuen.

Moravcsik, Andrew. 1993. "Integrating International and Domestic Theories of International Bargaining." In *Double-Edged Diplomacy: International Bargaining and Domestic Politics*, ed. Peter Evans, Harold Jacobson, and Robert Putnam, 3-42. Berkeley: University of California Press.

Nemeth, Tammy. 2001. "Continental Drift: Energy Policy and Canadian-American Relations." In *Diplomatic Departures: The Conservative Era in Canadian Foreign Policy, 1984-93*, ed. Nelson Michaud and Kim Richard Nossal, 59-70. Vancouver: University of British Columbia Press.

Neufeld, E.P. 1966. "Some Qualifying Thoughts on the CDC." *The Canadian Banker* 73: 29-33.

Nixon, Richard. 1972. Address of President Nixon to members of the Senate and of the House of Commons in the House of Commons Chamber. *House of Commons Debates*. Ottawa: House of Commons, April 14.

Noisi, Jorge. 1985. "Continental Nationalism: The Strategy of the Canadian Bourgeoisie." In *The Structure of the Canadian Capitalist Class*, ed. Robert Brym, 53-65. Montreal: Garamond Press.

Norrie, K.H. 1979. "Regional Economic Conflicts in Canada: Their Significance for An Industrial Strategy. In *Politics of an Industrial Strategy: A Seminar*, ed. Michael Jenkin, 55-83. Ottawa: Minister of Supply and Services.

Nossal, Kim Richard. 1985. "Economic Nationalism and Continental Integration." In *The Politics of Canada's Economic Relationship with the United States*, ed. Denis Stairs and Gilbert R. Winham, 55-96. Toronto: University of Toronto Press.

———. 1992. "Un pay européen? L'histoire de l'antlantisme au Canada." In *Making a Difference? Canada's Foreign Policy in a Changing World Order*, ed. John English and Norman Hillmer, 131-160. Toronto: Lester Publisher Limited.

———. 2001. "Bilateral Free Trade with the United States: Lessons from Canada." *Policy, Organization and Society* 20 (1): 47-62.

Nye, Joseph. 1988. "Neorealism and Neoliberalism." *World Politics* 40 (2): 235-251.

Organization for Economic Cooperation and Development. 1987. *Economic Survey: Canada*. Paris: OECD.

Ontario. Minister of Industry and Tourism. 1980. *The Report of the Advisory Committee on Global Product Mandating*, December.

———. 1981. *Interprovincial Economic Cooperation: Towards the Development of a Canadian Common Market*, January.

Painter, Martin. 1991. "Intergovernmental Relations in Canada: An Institutional Analysis." *Canadian Journal of Political Science* 24 (2): 269-88.

Pastor, Robert A. 2008. "The Future of North America: Replacing a Bad Neighbor Policy." *Foreign Affairs* 87 (4): 84-98.

Pearson, Lester. 1963. "Provision for Establishment of Department of Industry." *House of Commons*. Ottawa, June 7.

Pentland, C.C. 1982. "Domestic and External Dimensions of Economic Policy: Canada's Third Option." In *Economic Issues and the Atlantic Community*, ed. Wolfram Hanrieder, 139-62. New York: Praeger Publishers.

Pierson, Paul. 2000a. "Not Just What, but When: Timing and Sequence in Political Processes." *Studies in American Political Development* 14: 72-92.

———. 2000b. "Increasing Returns, Path Dependence, and the Study of Politics." *American Political Science Review* 94 (2): 251-67.

Pratt, Larry. 1982. "Energy: The Roots of National Policy." *Studies in Political Economy* 7: 27-59.

Phidd, Richard, and G. Bruce Doern. 1978. *The Politics and Management of Canadian Economic Policy*. Toronto: Macmillan Company of Canada Limited.

Quebec. 1973. Brief Submitted to the Standing Committee on Finance, Trade and Economic Affairs of the House of Commons, Concerning Bill C-132. 29th Parliament of Canada. Ottawa, June 19.

———. 1979. *Bâtir le Québec : un énoncé de politique économique*. Quebec: Ministère de l'industrie et du commerce.

Quebec, Minister Responsible for Privatization. 1986. *From the Quiet Revolution...To the Twenty-First Century: Report of the Committee on the Privatization of Crown Corporations*. Quebec, June.

Reisman, Simon. 2000. "The Negotiation and Approval of the FTA." In *Free Trade: Risks & Rewards*, ed. L. Ian MacDonald, 77-82. Montreal and Kingston: McGill-Queen's University Press.

Rose, Gideon. 1998. "Neoclassical Realism and Theories of Foreign Policy." *World Politics* 51 (1): 144-172.

Roy, Jacques. 1984. "The Tokyo Round: Canadian Perspectives." In *Canada/U.S. Trade Relations*, ed. Lee H. Radebaugh and Earl H. Fry, 34-42. Provo, Utah: Brigham Young University.

Ryan, Leo. 1974. "Resources are Trudeau's lever for trade agreement with EEC." *Globe and Mail* (October 26): B2.

Savoie, Donald. 2000. *Governing from the Centre: The Concentration of Power in Canadian Politics*. Toronto: University of Toronto Press.

Schultz, Richard, Frank Swedlove, and Katherine Swinton. 1980. *The Cabinet as a Regulatory Body: The Case of the Foreign Investment Review Act*. Ottawa: Economic Council of Canada.

Schultz, Richard. 1988. "Teleglobe Canada." In *Privatization, Public Policy and Public Corporations in Canada*, ed. Allan Tupper and G. Bruce Doern, 329-62. Halifax: Institute for Research on Public Policy.

Sharp, Mitchell. 1972. "Canada-U.S. Relations: Options for the Future." *International Perspective* (Autumn): 1-24.

———. 1981. "The Role of the Mandarins." *Policy Options* 2 (2): 43-4.

———. 1995. "The Trading Revolution." O.D. Skelton Memorial Lecture, Department of Foreign Affairs and International Trade, Alberta, February 7.

Shonfield, Andrew. 1970. *Modern Capitalism: The Changing Balance of Public and Private Power*. New York: Oxford University Press.

Simeon, Richard. 1979. "Federalism and the Politics of a National Strategy." In *The Politics of an Industrial Strategy: A Seminar*, ed. Michael Jenkin, 5-43. Ottawa: Science Council of Canada.

———. 1986. "Considerations on Centralization and Decentralization." *Canadian Public Administration* 29 (3): 445-461.

Simeon, Richard and Ian Robinson. 1985. *State, Society, and the Development of Canadian Federalism*. Toronto: University of Toronto Press.

Skogstad, Grace. 2002. "International Trade Policy and Canadian Federalism: A Constructive Tension?" In *Canadian Federalism: Performance, Effectiveness, and Legitimacy*, ed. Herman Bakvis and Grace Skogstad, 159-77. Don Mill, ON: Oxford University Press.

Smiley, Donald. 1975. "Canada and the Quest for a National Policy." *Canadian Journal of Political Science* 8 (1): 40-62.

————. 1977. "Territorialism and Canadian Political Institution." *Canadian Public Policy* 3 (4): 449-57.

Stairs, Denis. 1986. "Canada's Trade Relations with the United States: The Non-Economic Implications of an Economic Issue." In *Canada/U.S. Free Trade Agreement*, ed. Earl Fry and Lee Radebaugh, 46-69. Provo, Utah: Brigham Young University.

————. 1994. "Choosing Multilateralism: Canada's Experience After World War II and Canada in the New International Environment." *CANCAPS Papier* 4.

————. 1995. "Change in the Management of Canada-United States Relations in the Post-War Era." In *Toward a North American Community? Canada, the United States, and Mexico*, ed. Donald Barry, 53-74. Boulder, CO: Westview Press.

————. 1999. "The Pursuit of Economic Architecture by Diplomatic Means: The Case of Canada in Europe." In *Regionalism, Multilateralism and the Politics of Global Trade*, ed. Donald Barry and Ronald Keith, 228-52. Vancouver: University of British Columbia Press.

Stanbury, W.T. 1988. "Privatization and the Mulroney Government, 1984-1988. In *Canada Under Mulroney*, ed. Andrew Gollner and Daniel Salée, 119-57. Montreal: Véhicle Press.

Stanley, Guy, and Stephen Blank. 1999. "Big Gains for Canada." *The Gazette*, April 17.

Statistics Canada. 1986. *Canada's International Investment Position, 1981-1984*. Ottawa: Ministry of Supply and Services Canada.

————. 1991. *Canada's International Investment Position, 1988-1990*. Ottawa: Ministry of Supply and Services Canada, 1991.

————. 2007. *A Profile of Canadian Exporters, 1993 to 2005*. Catalogue no. 65-506-XIE. Ottawa: Ministry of Industry.

Stone, Frank. 1984. *Canada, the GATT and the International Trade System*. Ottawa: The Institute for Research on Public Policy.

Streeck, Wolfgang. 2001. "Introduction: Explorations into the Origins of Nonliberal Capitalism in Germany and Japan." In *The Origins of Nonliberal Capitalism: Germany and Japan in Comparison*, ed. Wolfgang Streeck and Kozo Yamamura, 1-38. Ithaca, NY: Cornell University Press.

Streeck, Wolfgang and Kathleen Thelen. 2005. "Introduction: Institutional Change in Advanced Political Economies." In *Beyond Continuity: Institutional Change in Advanced Political Economies*, ed. Wolfgang Streeck and Kathleen Thelen, 1-39. New York: Oxford University Press.

Thelen, Kathleen. 1999. "Historical Institutionalism in Comparative Politics." *Annual Review of Political Science* 2: 369-404.

Von Riekhoff, Harald. 1972. "The Recent Evolution of Canadian Foreign Policy: Adopt, Adapt and Improve." *The Round Table* 62: 63-76.

————. 1978. "The Third Option in Canadian Foreign Policy." In *Canada's Foreign Policy: Analysis and Trends*, ed. Brian Tomlin, 87-110. Toronto: Methuen.

Waldmann, Raymond. 1982. Statement of Raymond Waldmann before U.S. Senate, Committee on Foreign Relations, U.S. Economic Relations with Canada. 97th Congress of the United States. Washington DC, March 10.

Waltz, Kenneth. 1979. *Theory of International Politics*. Reading, MA: Addison-Wesley.

Weingast, Barry. 1995. "The Economic Role of Political Institutions: Market-Preserving Federalism and Economic Development." *Journal of Law, Economic, and Organization* 11 (1): 1-31.

Weiss, Linda. 1998. *The Myth of the Powerless State*. Ithaca, NY: Cornell University Press.

Westell, Anthony. 1984. "Economic integration with the USA." *International Perspective* (November/December): 5-22.

Weymen, Stephen. 1972. Statement before the Standing Committee on Finance, Trade and Economic Affairs of the House of Commons. 28th Parliament of Canada. Ottawa, June 13.

Whittington, Les. 1980. "Herb Gray's plea to cabinet." *Financial Times of Canada.* 69 (September 15): 1-2.

Williams, Glen. 1987. "Canadian Sovereignty and the Free Trade Debate." In *Knocking on the Back Door: Canadian Perspectives on the Political Economy of Freer Trade with the United States*, ed. Allan Maslove and Stanley Winer, 101-19. Montreal: Institute for Research on Public Policy.

———. 1994. *Not For Export: Toward a Political Economy of Canada's Arrested Industrialization.* Toronto: McClelland and Stewart Limited.

Wilson, Michael. 2000. "Free Trade: Then and Now." In *Free Trade: Risks & Rewards*, ed. L. Ian MacDonald, 207-10. Montreal and Kingston: McGill-Queen's University Press.

Winham, Gilbert. 1978-9. "Bureaucratic politics and Canadian trade negotiation." *International Journal* 35 (1): 64-89.

———. 1986. *International Trade and the Tokyo Round Negotiation.* Princeton, NJ: Princeton University Press.

———. 1994. "NAFTA and the trade policy revolution of the 1980s: a Canadian perspective." *International Journal* 49: 472-508.

Winham, Gilbert, and Elizabeth DeBoer-Ashworth. 2000. "Asymmetry in Negotiating the Canada-US Free Trade Agreement, 1985-1987. In *Power and Negotiation*, ed. William Zartman and Jeffrey Rubin, 35-52. Ann Arbor: The University of Michigan Press.

Winham, Gilbert, and Sylvia Ostry. 2003. "The second trade crisis." *Globe and Mail*, June 17.

Wolfe, David. 1978. "Economic Growth and Foreign Investment: A Perspective on Canadian Economic Policy, 1945-1957." *Journal of Canadian Studies* 13: 3-20.

———. 1984. "The Rise and Demise of the Keynesian Era in Canada: Economic Policy, 1930-1982." In *Modern Canada, 1930-1980's*, ed. Michael Cross and Gregory Kealey, 46-80. Toronto: McClelland and Stewart.

Wonnacott, Ronald. 1990. "U.S. Hub-and-Spoke Bilateral and the Multinational Trading System." *C.D. Howe Institute Commentary* 23.

Wood, Stewart. 2001. "Labour Market Regimes under Threat? Sources of Continuity in Germany, Britain, and Sweden." In *The New Politics of the Welfare State*, ed. Paul Pierson, 368-409. New York: Oxford University Press.

———. 2001. "Business, Government, and Patterns of Labor Market Policy in Britain and the Federal Republic of Germany." In *Varieties of Capitalism: The Institutional Foundations of Comparative Advantage*, ed. Peter A. Hall and David Soskice, 247-74. New York: Oxford University Press.

Wright, Gerald, and Maureen Appel Molot. 1974. "Capital Movements and Government Control." *International Organization* 28 (4): 671-88.

Yarbrough, Beth, and Robert M. Yarbrough. 1992. *Cooperation and Governance in International Trade: The Strategic Organizational Approach.* Princeton, NJ: Princeton University Press.

Yeutter, Clayton. 2000. "The Negotiation and Approval of the FTA." In *Free Trade: Risks & Rewards*, ed. L. Ian MacDonald, 74-7. Montreal and Kingston: McGill-Queen's University Press.

Young, R.A, Philippe Faucher, and André Blais. 1984. "The Concept of Province-Building: A Critique." *Canadian Journal of Political Science* 17 (4): 783-818.

Zysman, John. 1983. *Governments, Markets, and Growth: Financial Systems and the Politics of Industrial Change*. Ithaca, NY: Cornell University Press.

———. 1994. "How Institutions Create Historically Rooted Trajectories of Growth." *Industrial and Corporate Change* 4 (1): 243-83.

Index